Working with Semiconductors

Working with Semiconductors

Albert C, W, Saunders

With a specially written chapter for the guidance of the English reader by W. Oliver

FOULSHAM-TAB LIMITED
Slough Bucks England

Foulsham-Tab Limited
Yeovil Road Slough Bucks England

Working with Semiconductors

Library of Congress Number 78–85323
Cat. Code No. 501

ISBN 0–7042–0114–3

Introduction Printed and Made in Great Britain by
A. Wheaton & Co., Exeter

Balance printed in U.S.A.

Nowadays, if you are involved in any way with electronics, you are sure to find yourself constantly working with semiconductors. In the present solid-state era, these devices have come into such universal use for such an enormous variety of applications that you simply cannot get away from them, and in the circumstances, books on the subject of this one are essential reading for anyone who takes any active interest in electronics at either an amateur or professional level.

Although most of the text is concerned mainly with discrete semiconductors—i.e., separate diodes, transistors, FET's and so on—the principles it discusses are largely applicable also to the more recent development that has stemmed from these devices; namely, the IC or integrated circuit, which is discussed in Chapter 7.

This book originates from the United States, and since it was written the use of IC's both in America and in Britain (as well as in other countries) has increased enormously. They are now widely available, usually at quite reasonable prices when one considers the number of discrete devices and components they replace; and they are being used now in a very wide variety of applications.

Where specific semiconductors are referred to in this book they are of course American types. Many American diodes, transistors, FET's etc. are now listed by suppliers in Britain; and there are also British comparable types which can be substituted in many cases for American ones specified.

A few examples of comparables (among types mentioned in these pages) are: American 2N217 and British AC128; 2N301 and AD149; 2N1102 and AC127; 2N2926 and BC107; 2N35 and AC127; 2N265 and AC128; etc. You will see from the foregoing that some British types will substitute for several American ones.

Similarly with diodes, for the 1N64 you can substitute a British OA90 or AA119; for a 1N541 you can substitute an AA119; and for a 1N87 you can, again, use an OA90.

According to makers' numberings there would appear to be many thousands of different transistors. But in reality there are hundreds of cases in which different manufacturers have given

v

different numbers to what are virtually the same transistors, so the variety is not nearly as great as it appears to be; not that this makes the task of transistor selection any less confusing!

Substitution lists are available from a good many sources —makers, distributors and book publishers—so it is not too difficult to pick suitable alternatives for many types that happen to be hard to get, or obsolete.

New types are constantly appearing, and the exacting requirements of military, industrial and scientific users mean that they discard hundreds of types as obsolete years before the ordinary amateur or hobbyist gives up using them. Everything depends on one's individual requirements and if a theoretically or officially "obsolete" type will meet those requirements perfectly then it may be quite a good buy, if you can get it very cheaply, as often happens.

But if there is not much difference in price between an older type and a latest one, it is usually by far the best plan to get the updated type as it will stay "fashionable" much longer. Anyone who wants to experiment with a lot of different constructional projects may wish to re-use a given transistor over and over again in successive applications.

If you handle the transistor carefully enough, repeated re-use is quite feasible. But if you plan to do this, it is best to use a plug-in transistor holder or some equivalent means of making connection, because continually heating the lead-out wires in soldering and unsoldering is likely to shorten the life of the transistor, though this risk can be minimized admittedly if you avoid cutting short the wires, use insulating plastic sleeving on them, and grip them with fine-nosed pliers as a heat-sink when applying the soldering-iron.

On p. 103 the American author refers to an IF of 455 kHz. This is a standard choice of IF in the majority of American transistor receivers, and is now a favourite one also in British and foreign-made sets. But earlier British models were apt to be designed with other IF's—in fact, years ago in the days of valve sets the number of different IF's used by various British manufacturers made constant reference to data-sheets an essential task when servicing and re-aligning a wide variety of receivers. But nowadays you are fairly safe in assuming an IF of 455 kHz in ordinary AM circuits. VHF/FM sets and car radios use different IF's.

In the chapter on power supplies (Chapter 9) you will note

references to the 117 volts AC and 60 Hz frequency which are typical of American domestic mains supplies, in contrast to our 240 VAC at 50 Hz. Where a double-wound mains transformer is used (and this is one of the safest methods of isolating equipment from risk of direct contact with the mains-supply leads), the primary must be a British standard winding suitable for our 240-volt mains, instead of the American type designed for their 110–120-volt supplies.

When using mains supplies for any purpose whatsoever it is essential to make sure that all the equipment is of adequately shock-proof design and that all necessary safety precautions are strictly observed. Great danger can arise from attempting to use transistor equipment designed for low-voltage batteries on any sort of mains-powered battery-eliminator unit unless adequate safeguards have been provided to prevent any possible risk of exposed metalwork on the transistor set becoming "live" to the mains.

Semiconductors seem to be a particularly favourite form of merchandise for a great many mail-order and retail radio suppliers nowadays. These small items are easy to handle, pack and post, which is probably one reason for their popularity in this direction. Some types are very plentiful, but others are in short supply and you may experience a lot of delay in getting them. If alternatives will do it is best to state this when ordering.

A look through the advertisement pages of the leading technical journals will give you the names and addresses of a great many firms handling semiconductors and other solid-state products, such as integrated circuits.

Regarding other components, you may find it difficult or impossible to get some American types specified in this or other books of American origin, and you will need to substitute British-made alternatives. In finding these, it is a great help to have comprehensive catalogues which can be bought from some of the principal suppliers that specialize in catering for the radio, television and electronic enthusiasts.

Preface

Since the first experimental transistors appeared in the Bell Telephone Laboratories two decades ago, the field of semiconductors has grown—quite literally—by leaps and bounds, making possible the application of electronic wonders in fields where older tube-type circuits would have been ruled out solely on overall size and power requirements alone.

Of course, the transistor was just the beginning—the forerunner, you might say—of an increasingly vast array of semiconductor devices, beginning with the lowly germanium diode. Now, the field of semiconductors has grown to the point where the size of an entire circuit is smaller than most transistors! And who knows what lies just over the horizon?

Within these pages you'll learn all about the various semiconductor devices in use today. Beginning with just enough background data on semiconductor technology, you will be exposed to the make-up of diodes, transistors, and then components combining the attributes and characteristics of both. Most of the book, though, deals with applications, all with a strong practical link to modern electronics.

The material included in this book is designed to serve as a study guide as well as a collection of experiment-type "projects." Component values are included in most circuits to enable you to actually construct or "breadboard" those which interest you most. As you'll discover (if you haven't already) working with semiconductors can be fun, as well as educational. My purpose will be accomplished if this book serves you in either, or both, of these aspects.

Albert C. W. Saunders

CONTENTS

Chapter 1

Getting Acquainted with Semiconductors

A semiconductor, as the name implies, is neither a good conductor nor a good insulator. It has a resistivity somewhere between metallic and dielectric materials. The electrical characteristics of semiconductor devices are similar to the forerunner—the crystal diode; therefore, a brief description of the diode will be helpful in the study of other semiconductor developments.

CRYSTAL DIODE

The operation of a crystal diode depends upon the conductivity of such substances as germanium, silicon, selenium, etc. Germanium and silicon, being the most applicable of the group, are widely used in construction of diode and transistor devices. Either substance, in its purest form, is an insulator and, therefore, useless for diode or transistor construction. However, when an impurity is added to either germanium or silicon, each acquires certain electronic characteristics and becomes a semiconductor.

The mechanical construction of semiconductor devices falls into two major classes—point contact and junction. The latter was selected to introduce and simplify the study of semiconductor phenomena.

JUNCTION DIODE

The heart of a crystal diode or transistor is a junction formed by two types of impure germanium pellets, "negative" or "N" type and "positive" or "P" type. Independently, both

9

pellets conduct equally in either direction when the applied potential is AC, but when the two pellets are joined to form a junction, the device becomes a rectifier. See Fig. 1-1. The change from a bi-directional to a uni-directional current flow is attributed to the dissimilar electrical characteristics of the "N" and "P" type germanium pellets that form the junction. To understand how these pellets are processed to acquire "N" and "P" type properties, a brief summary of atomic structure is necessary. This is not difficult to understand, since it explains the difference in the physical properties and electrical behavior of semiconductors.

ATOMIC STRUCTURE

An atom is the smallest unit particle of matter that cannot be subdivided without losing its identity. The atom contains energy particles called protons (positive electricity), neutrons (neutral, except hydrogen which has only one proton and one electron) and electrons (negative electricity).

The protons and neutrons form the nucleus or core of the atom around which revolve electrons in regular orbits, like planets around a central sun. See Fig. 1-2. In an electrically-balanced atom there are as many electrons (negative charges), as there are protons (positive charges), and the number of protons associated with an atom is the same as its atomic number. The Table in Fig. 1-2 lists a few elements taken from the chart of known elements.

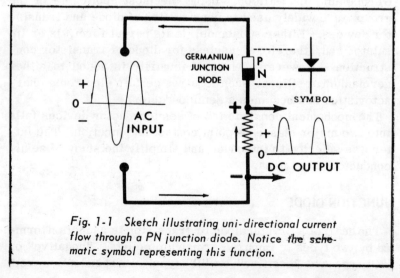

Fig. 1-1 Sketch illustrating uni-directional current flow through a PN junction diode. Notice the schematic symbol representing this function.

Element	Atomic No.	Element	Atomic No.
Hydrogen	1	Aluminum	13
Helium	2	Silicon	14
Carbon	6	Germanium	32
Oxygen	8	Arsenic	33
Neon	10	Selenium	34
Sodium	11	Indium	49

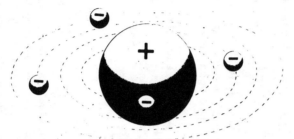

Fig. 1-2. This drawing illustrates the orbital path of electrons around the nucleus of an atom. Such orbits are generally referred to as "shells or rings" in atomic theory.

Each atom has a certain number of orbits and in each orbit there are a certain number of electrons. However, it is not the total number of electrons in the atom that has any significance in this study but the normal number of electrons in its outermost orbit or shell. See Fig. 1-3. These are called "valence electrons" and are Nature's knitting needles used to bond atoms together. They coordinate in "pairs" and "octets" with the valence electrons of neighboring atoms, as will be illustrated later.

ELECTRONIC THEORY OF VALENCE

This theory deals with atomic structures and particularly with the arrangement of valence electrons in the outer shell. It states that an octet (8 electrons) is stable and is formed by the sharing of electrons between the atoms of a substance.

Germanium, arsenic, and indium are some of the substances used in processing "N" and "P" type germanium and the num-

11

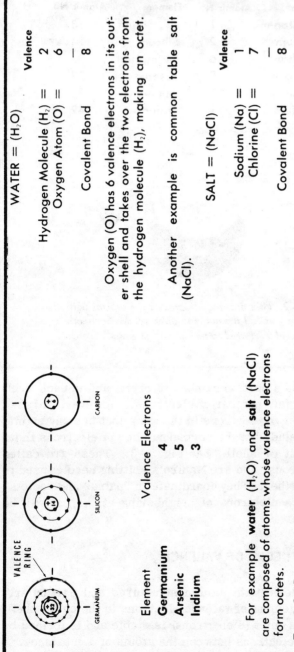

WATER = (H_2O)

		Valence
Hydrogen Molecule (H_2)	=	2
Oxygen Atom (O)	=	6
Covalent Bond		8

Oxygen (O) has 6 valence electrons in its outer shell and takes over the two electrons from the hydrogen molecule (H_2), making an octet.

Another example is common table salt (NaCl).

SALT = (NaCl)

		Valence
Sodium (Na)	=	1
Chlorine (Cl)	=	7
Covalent Bond		8

VALENCE RING

GERMANIUM   SILICON   CARBON

Element	Valence Electrons
Germanium	4
Arsenic	5
Indium	3

For example water (H_2O) and salt (NaCl) are composed of atoms whose valence electrons form octets.

Fig. 1-3. Diagrams representing germanium, silicon, and carbon atoms to illustrate the arrangement of electron orbits or rings surrounding the nucleus. The GERMANIUM ATOM has four rings each containing 2-8-18-4 electrons respectively. The SILICON ATOM has three rings each containing 2-8-4 electrons respectively. The CARBON ATOM has two rings each containing 2-4 electrons respectively. The valence electrons associated with each atom are shown in the outer ring. Each valence electron has been given a minus sign for the purpose of illustrating covalent bonds. See Fig. 1-4.

ber of valence electrons associated with each substance appears in Fig. 1-3.

COVALENT BONDS

The diamond,[*] which is a stable formation of carbon atoms, serves as an excellent illustration of how atoms can be tightly knit together. The carbon atom has four valence electrons which coordinate their movements with the valence electrons of neighboring atoms. Together they form "covalent bonds" consisting of four pairs or an octet that establishes an equilibrium whereby the loosely bound valence electrons are then tightly bound to the nuclei. See Fig. 1-4A. Carbon is a semiconductor but in the form of a diamond it is a very hard substance and is an insulator. Germanium, like the carbon atom, has four valence electrons that coordinate to form covalent bonds which tightly knit the atoms together, forming a crystal, thus making it an insulator.

SEMICONDUCTOR "N" TYPE

When two elements having a different number of valence electrons are combined, a similar coordination between the valence electrons takes place. For example, combining arsenic which has five valence electrons with germanium which has four, the two groups coordinate in pairs to form an octet, thus tightly binding eight of nine valence electrons to the nucleus. This leaves one electron mobile and free, thus making the germanium crystal a "negative" or "N" type semiconductor. The conductivity of the crystal depends upon the number of free electrons available, which is governed by the amount of impurity added. One arsenic atom to every 100 million germanium atoms is adequate for transistor operation. Connecting a battery across an "N" type crystal causes the free electrons to flow through the crystal from the negative to the positive terminal. Therefore, the major current carriers in the "N" type germanium crystal are electrons. See Fig. 1-4B.

SEMICONDUCTOR "P" TYPE

Elements used for processing the "P" type crystal are ger-

---

*Scientists have recently taken over this product of Nature by producing artifical diamonds. A similar claim was made by Moissen in 1893.

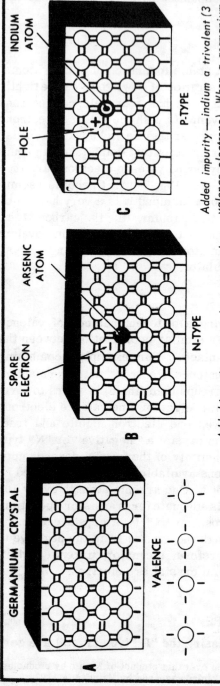

GERMANIUM CRYSTAL

A

VALENCE

Covalent bond structure in a tetravalent (4 valence electrons) atom lattice, such as germanium, silicon, and carbon, the latter being in the form of a diamond. When valence electrons are involved in covalent bonds, they are not free and cannot take part in the conduction of electricity.

B

SPARE ELECTRON

ARSENIC ATOM

N-TYPE

Added impurity—arsenic a pentavalent (5 valence electrons). When a germanium and arsenic atom are involved in covalent lattice structure there is one electron left over. This electron readily enters the conduction band and acts as a free electron or current carrier. In this case the arsenic atom is called the DONOR ATOM.

C

HOLE

INDIUM ATOM

P-TYPE

Added impurity—indium a trivalent (3 valence electrons). When a germanium and indium atom are involved in covalent bonds there is one electron short. The indium atom acquires an additional electron from a nearby germanium atom and forms covalent bonds with neighboring germanium atoms. The covalent band that was robbed of an electron acquired a hole. A hole is a mobile energy particle which readily acts as a carrier of positive electricity. In this case the indium atom is called an ACCEPTOR ATOM.

Fig. 1-4.

14

manium and indium. The latter has three valence electrons. When germanium and indium atoms are combined, each indium atom will remove an electron from a nearby germanium atom, leaving a "hole" (positive charge) in the germanium crystal. The electron accepted by the indium atom provides the required four valence electrons which tightly knit it to neighboring germanium atoms whose four valence electrons are left intact, thereby forming an octet. Remember, there is only one impurity atom added to every 100 million germanium atoms; therefore, not all germanium atoms are involved in the exchange and only a few of them acquire a positive charge. However, the small amount of impurity added provides a sufficient number of positive charges (holes) for transistor operation. A hole represents a deficit of one electron, and constitutes a current carrier simulating a mobile electron but with a positive charge. See Fig. 1-4C. It is possible to use other impurities which possess three valence electrons without changing the symmetry of the crystal, such trivalents as boron, galium, or aluminum.

When a battery is applied across the "P" type crystal, the holes move through the crystal from the positive to the negative terminal. Each hole on arriving at the negative terminal will receive an electron, while at the positive terminal a covalent bond is broken down and an electron is released to the battery. Hence, the major current carriers in the "P" type crystal are holes or positive charges. Notice that the terms "hole" and "positive charge" are interchangeable in the text.

POSITIVE CARRIER OR HOLE CONDUCTOR

The movement of electrons through a conductor is fairly well understood. However, the movement of "holes" (positive carriers) requires some additional explanation. The action can be demonstrated by a simple experiment using an electroscope, an instrument consisting of a metal rod suspended by an insulator; it has a ball terminal on top and two strips of gold leaf suspended at the bottom. See Fig. 1-5.

When a positively-charged object is held near the ball terminal, free electrons in the rod are attracted to it. Each electron leaves behind a hole (positive charge). The holes migrate in the opposite direction and congregate on the two strips of gold leaf. Since the holes represent positive charges, the two strips repel one another causing them to diverge. Notice that the holes travel in a direction opposite to the electron flow and constitute the carriers of positive electricity. The presence of holes in this experiment is only temporary since the electrons are free to move back as soon as the charged rod is removed. In the "P" type germanium crystal the holes that are cancelled at the negative terminal of the battery are immediately replaced at the positive terminal, thus keeping the number of holes (positive charges) present in the materials at a constant level.

In semiconductor theory an accurate definition of a hole is: "The positive charge remaining when an electron is removed from a covalent bond." Merely removing an electron from a neutral atom does not create a true hole as those developed in "P" type germanium or silicon, where a special process is used.

Germanium for transistor work must be of perfect crystal-

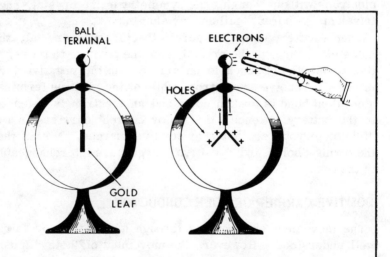

Fig. 1-5. Here is an experiment in electrostatics illustrating the existence and direction of movement of electrons and holes. In semiconductor theory an accurate definition of a hole is "the positive charge remaining when an electron is removed from a covalent bond." Merely removing an electron from a neutral atom does not create a true hole as those developed in P type germanium. In the true sense of the word—a hole is the absence of an electron that cannot return to fill the vacancy in a covalent bond.

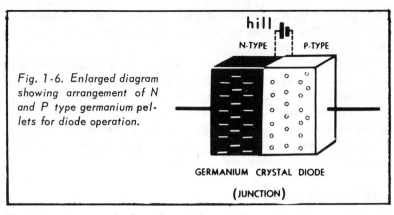

Fig. 1-6. Enlarged diagram showing arrangement of N and P type germanium pellets for diode operation.

GERMANIUM CRYSTAL DIODE

(JUNCTION)

line structure and of uniform chemical composition. Natural germanium contains many impurities and does not meet these requirements. From a "melt" made of the natural product, the impurities are separated; then selected impurities are added to produce the "N" and "P" type crystals.

The number of impurity atoms determines the conductivity of a germanium crystal. Adding an impurity that is in the range of 0.5 to 8 parts in 100 million provides the proper conductivity for transistors. To obtain the proper ratio the germanium must be purified to reduce impurities to approximately one part in a billion prior to the injection of the selected impurity. Since the impurity atoms electronically replace germanium atoms, the symmetry of the crystal structure is not destroyed and an ideal semiconductor crystal is produced.

The extraction and refining process of germanium costs about $225 per pound but since a transistor requires very tiny pellets the cost is not prohibitive.

SEMICONDUCTOR JUNCTION

Joining small pellets of "N" and "P" germanium together as shown in Fig. 1-6 produces an effective diode rectifier. In making this junction it would seem that the mobile electrons in the "N" pellet would migrate to the "P" pellet and cancel the holes. However, this is not so; instead, they repel one another and establish a potential difference of a few tenths of a volt across the junction. This is easy to understand if we regard the arsenic and indium impurities as the positive and negative plates of a battery. Since the arsenic atom in the

"N" pellet <u>lost</u> an electron it acquired a positive charge. In the "P" pellet the indium atom <u>gained</u> an electron and acquired a negative charge.

In elementary electronics it was learned that when a neutral atom lost an electron it became a positive ion; when it gained an electron it became a negative ion. In the processing of "N" and "P" type crystals the atoms of both impurities become and remain ions; arsenic a positive, indium a negative. In semiconductor terminology the arsenic atom is called a "donor" and the indium an "acceptor."

"POTENTIAL HILL"

The negatively-charged impurity atoms in the "P" pellet repel the mobile electrons in the "N" pellet and vice versa. Therefore, the two impurities resemble the plates of a small battery immersed in germanium and will not short-circuit

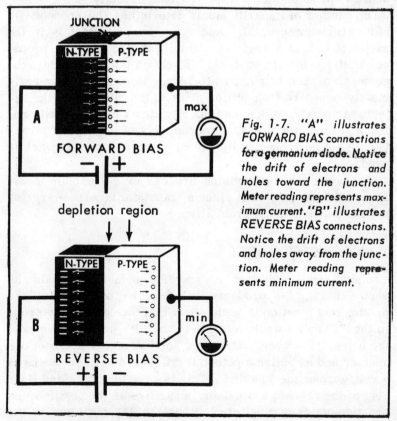

Fig. 1-7. "A" illustrates FORWARD BIAS connections for a germanium diode. Notice the drift of electrons and holes toward the junction. Meter reading represents maximum current. "B" illustrates REVERSE BIAS connections. Notice the drift of electrons and holes away from the junction. Meter reading represents minimum current.

internally. When this "potential hill" is broken down by an external voltage, sufficient current flows through the junction.

FORWARD AND REVERSE BIAS

When an external bias voltage is applied to a crystal diode, the amount of current flow depends upon the applied polarity.

The diagrams in Fig. 1-7A and 1-7B illustrate FORWARD and REVERSE bias connections. Fig. 1-7A is an example of FORWARD BIAS. The negative terminal of the battery connected to the "N" type material repels the free electrons toward the junction where "N" touches "P." The positive terminal connected to the "P" type material repels the holes toward the junction where "P" touches "N." This causes the inherent potential barrier to break down, resulting in a comparatively large current due to the combining of electrons and holes at the junction.

Fig. 1-7B is an example of REVERSE BIAS. The positive terminal connected to the "N" type material attracts the free electrons, and the negative terminal connected to the "P" type material attracts the holes. This concentrates electrons and holes away from the junction which increases the inherent barrier potential with the result that no appreciable current can flow through the junction. Such a barrier is called the "depletive area" and is capacitive in nature.

Summarizing, we find that the FORWARD BIAS broke down the inherent potential barrier and permitted the exchange of free electrons and holes, thereby developing a low resistance path to the flow of current. The REVERSE BIAS increased the inherent potential barrier, thereby developing a high resistance path. Despite the reverse bias connection there will be a small current flow due to stray carriers in the vicinity of the junction. The uni-directional current characteristic of the crystal diode has made it a valuable component in many circuits were "RF" and "IF" detection is required. It also is used extensively as a rectifier in DC power supplies, and has many other significant applications yet to be discussed.

THE TRANSISTOR

The word "transistor" is coined from the term TRANSfer-resISTOR. A TRANSISTOR is a semiconductor device capable of amplification and many other functions now performed by vacuum tubes. Its compactness, low power consumption, and amplification properites offer unlimited possibilities in the design of lightweight electronic equipment. The transistor requires <u>two</u> junctions for its operation and is merely an extension of the diode.

NPN JUNCTION TRANSISTOR

We have learned that a JUNCTION of "N" and "P" type germanium forms a crystal diode, so let's consider the construction and operation of an NPN JUNCTION TRANSISTOR. If two germanium diodes were connect in series so that the "P" type sections were alloyed together, basically we would have an "NPN" junction transistor. However, in actual construction the center is a thin layer of "P" type germanium sandwiched between two layers of "N" type germanium. See Fig. 1-8A. In manufacture the three layers are fused together and designated EMITTER, BASE, and COLLECTOR, respectively. This provides a semiconductor device having two junctions. When the two junctions are correctly biased, electrons flow from the "N" type emitter to the collector. A

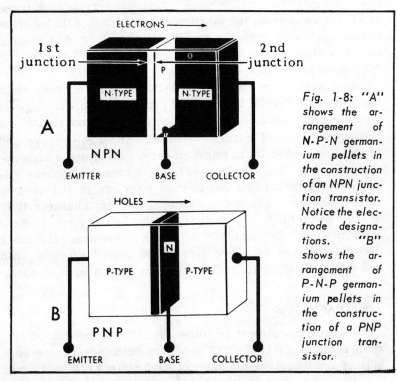

Fig. 1-8: "A" shows the arrangement of N-P-N germanium pellets in the construction of an NPN junction transistor. Notice the electrode designations. "B" shows the arrangement of P-N-P germanium pellets in the construction of a PNP junction transistor.

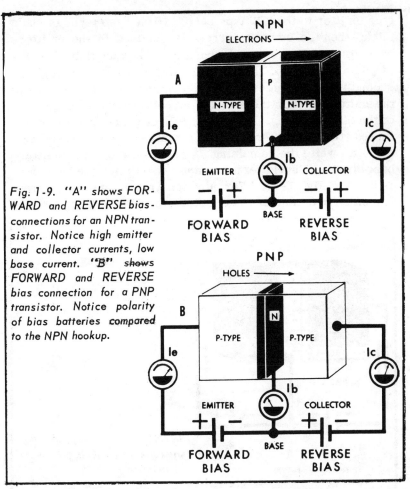

Fig. 1-9. "A" shows FOR-WARD and REVERSE bias-connections for an NPN transistor. Notice high emitter and collector currents, low base current. "B" shows FORWARD and REVERSE bias connection for a PNP transistor. Notice polarity of bias batteries compared to the NPN hookup.

few of the electrons combine with holes in the "P" base section, but the majority enter the "N" type collector.

Adding an extra junction to a germanium diode is similar to adding a control grid to a vacuum tube diode, since both devices acquire amplification properties. Notice that a FORWARD BIAS is applied to the EMITTER-BASE junction to provide a low resistance path and a REVERSE BIAS is applied to the COLLECTOR-BASE junction to provide a high resistance path. See Fig. 1-9A and 1-9B.

CURRENT TRANSFER

The applied emitter bias voltage overcomes the potential barrier of the first junction, and electrons pass from the "N" type emitter to the "P" type base. If the base layer is made thick, then the emitter current is confined to the emitter-base circuit. However, the base layer is very thin, and approximately 95% of the emitter current passes through the second junction, enters the collector circuit, and returns to the emitter through the external circuit under the compulsion of the bias batteries. Notice that 95% of the current is transferred from the low-resistance circuit to the high-resistance circuit, while the remaining 5% passes through the emitter-base circuit. This latter percentage constitutes the base current. See Fig. 1-10. For junction transistors the current transfer from emitter to collector may be as high as 99%.

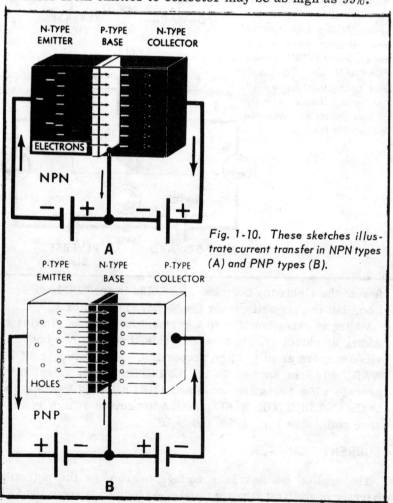

Fig. 1-10. These sketches illustrate current transfer in NPN types (A) and PNP types (B).

$$R_{GAIN} = \frac{R_{OUTPUT}}{R_{INPUT}} = \frac{900K}{500} = 1800$$

$$\text{VOLTAGE GAIN} = .95 \times \frac{900K}{500} = 1710$$

$$\text{POWER GAIN} = (.95)^2 \times \frac{900K}{500} = 1624$$

PNP JUNCTION TRANSISTOR

The "PNP" junction transistor has operating characteristics similar to the "NPN" junction transistor just discussed, but differs in the arrangement of the "N" and "P" electrodes as illustrated in Fig. 1-8B. Notice that the emitter and collector electrodes have been changed from "N" type to "P" type and the base from "P" type to "N" type. In order to provide forward and reverse bias to the emitter and collector circuits, respectively, the polarity of the bias batteries is reversed. Operation of the two transistors differs only in the method of current conduction as follows (see Fig. 1-10A and 1-10B):

1. The current carriers through the "NPN" junction transistors are ELECTRONS.
2. The current carriers through the "PNP" junction transistors are HOLES (positive carriers).

The collector current in both transistors is approximately 5% less than the emitter current; the ratio of output resistance (Ro) to input resistance (R1) is substantially the same in both types.

If the emitter current is 2 ma, the collector current will be 95% of 2 ma or 1.90 ma. Therefore, current gain α (alpha) of the junction transistor will be as follows:

$$\text{CURRENT GAIN} = \frac{I_C}{I_E} = \frac{1.9}{2} = .95$$
$$\text{(alpha)}$$

The current gain of less than "one" may seem disappointing. However, if we consider the low-resistance path of the emitter circuit and the high-resistance path of the collector circuit, and take into account, that both are conducting nearly equal current, there should be considerable voltage gain. Assum-

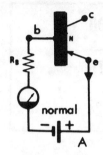

DIAGRAM A — NORMAL BASE BIAS

Biasing the base of a "PNP" transistor **Negatively** with respect to the emitter, the base-emitter junction will offer a relatively low resistance (**Forward Bias**), allowing a forward current to flow through the junction proportional to the applied bias voltage. This is the normal procedure for obtaining bias current.

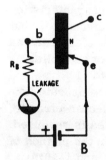

DIAGRAM B — INCORRECT BASE BIAS

Biasing the base ("PNP") **Positively** with respect to the emitter, the base-emitter junction will offer a relatively high resistance (**Reverse Bias**), a comparatively small leakage current will flow through the junction due to impurities. This type of bias connection would rectify the input signal and cause distortion.

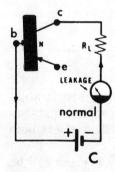

DIAGRAM C — NORMAL COLLECTOR BIAS

Biasing the collector of a "PNP" transistor **Negatively** with respect to the base, the base-collector junction will offer a relatively high resistance (**Reverse Bias**). A small leakage current will flow due to impurities. This is the normal procedure.

DIAGRAM D — INCORRECT COLLECTOR BIAS

Biasing the collector ("PNP") **Positively** with respect to the base, the base-collector junction will offer a relatively low resistance (**Forward Bias**), causing a relatively large current to flow that is proportional to the applied bias. This can cause damage to the transistor due to the higher collector voltage.

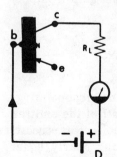

The arrows shown in the diagrams indicate direction of electron current.

For "NPN" transistors the applied bias voltages are reversed for normal operation.

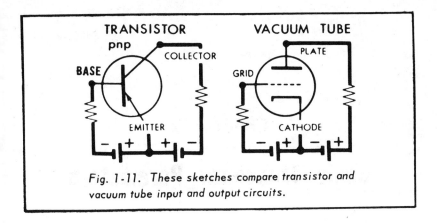

Fig. 1-11. These sketches compare transistor and vacuum tube input and output circuits.

ing the emitter input resistance to be 500 ohms and collector output resistance to be 900,000 ohms, the voltage gain (VG) of the junction transistor will be as indicated on the accompanying chart. Although the current gain is only .95, a considerable voltage and power gain is realized due solely to the ratio of output resistance (Ro) to input resistance (Ri).

TRANSISTOR AND VACUUM TUBE COMPARISONS

A vacuum tube amplifier normally operates with a fixed negative grid voltage. This constitutes a high-impedance input circuit. The fixed positive voltage applied to the plate constitutes a low-impedance output circuit. A transistor amplifier normally operates with forward bias applied to its base. This constitutes a low-impedance input circuit. The reverse bias applied to the collector constitutes a high-impedance output circuit. See Fig. 1-11.

Comparing these differences between the two units emphasizes the fact that a vacuum tube is voltage-controlled, while the transistor is current-controlled. In a vacuum tube current flows through a vacuum, while in a transistor current flows through a solid. In these comparisons, the term "impedance" has been used interchangeably with resistance, since only "DC" potentials are involved.

Chapter 2

Basic Transistor Parameters

The purpose of this study is to acquaint the reader with the practical version of transistor circuitry. The project consists of constructing a simple transistor checker that can measure LEAKAGE CURRENT (Iceo),* CURRENT GAIN ALPHA, and CURRENT GAIN BETA. The circuit diagram in Fig. 2-1 is biased for PNP transistor tests. To test NPN transistors, reverse the polarity of the battery and the meter leads. A protective resistor (R1) is connected in series with the emitter to prevent damage to the meter, should the transistor leads be shorted accidentally.

LEAKAGE TEST

For the simultaneous measurement of emitter, base, and collector currents, two additional 0-1 ma meters are required. See Fig. 2-2. The EMITTER-COLLECTOR circuit is closed by toggle switch SW 1. With SW 1 closed and SW 2 open the meter registers LEAKAGE CURRENT (Iceo)*. Notice with SW 2 open the BASE CURRENT is zero. See Fig. 2-1. The leakage current of the transistor under test measured .02 ma (20 microamperes).

INCREASED TEMPERATURE OBSERVATIONS

While the leakage test is in progress squeeze the transistor housing between the thumb and first finger and observe a small increase in leakage current due to body heat. Although the in-

---

*Iceo represents the current that flows from the emitter to collector with the base circuit open.

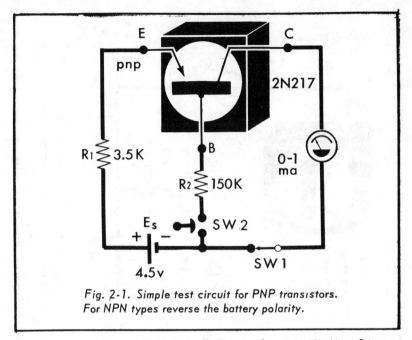

Fig. 2-1. Simple test circuit for PNP transistors. For NPN types reverse the battery polarity.

crease is very small it does indicate the sensitivity of transistors to temperature. When a hot soldering iron (90 watts) was held about one inch from the transistor, the leakage current gradually increased from a normal .02 ma to .2 ma, and was still rising when the iron was removed. The meter reading returned to normal (.02 ma) five minutes after the iron was moved away. An increase in leakage current due to thermal activity reduces transistor power output and increases distortion. That's why transistor characteristics and maximum ratings supplied by the manufacturer always specify ambient temperature.

CURRENT GAIN (BETA)

A change in base current also causes a change in collector current, and since the base current is about 5% of the emitter current, a gain greater than 1 is obtained. This gain factor is called beta (symbol β) and is expressed by the following ratio:

$$\frac{\text{Change in collector current}}{\text{Change in base current}}$$

27

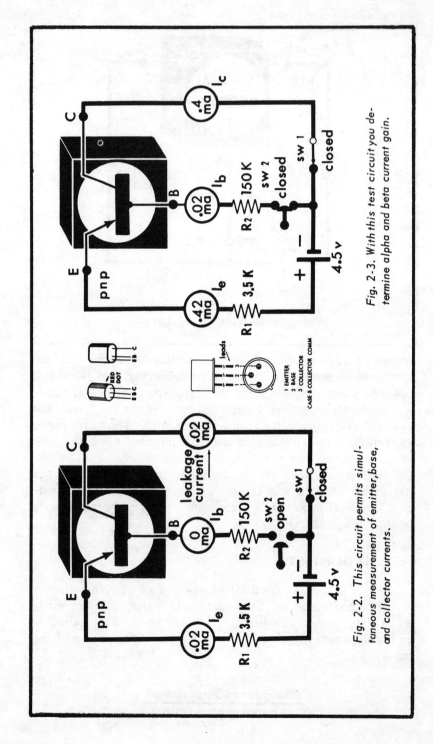

Fig. 2-3. With this test circuit you determine alpha and beta current gain.

Fig. 2-2. This circuit permits simultaneous measurement of emitter, base, and collector currents.

1 EMITTER
2 BASE
3 COLLECTOR
CASE & COLLECTOR COMM

and is expressed mathematically as follows:

$$\text{BETA} = \frac{\triangle I_c}{\triangle I_b}$$

For junction transistors beta is always greater than 1.

Refer to Fig. 2-2. The collector current is .02 ma and the base current is zero. Now refer to Fig. 2-3. It will be seen that when SW 2 is closed the base current rises from zero to .02 ma (20 microamperes). This causes the collector current to increase from .02 ma to .4 ma. It is interesting to note at this point that for a base current increase of .02 ma, the collector current increased 19 times; this is called the BETA AMPLIFICATION FACTOR, and may be expressed mathematically as follows:

$$\text{BETA} = \frac{.4\,ma - .02\,ma}{.02\,ma} = \frac{.38}{.02} = 19$$

The relation of BETA to ALPHA is expressed by the following equation:

$$\text{BETA} = \frac{\alpha}{1 - \alpha}$$

This equation indicates that the nearer ALPHA is to unity, the greater BETA becomes. Example: What is the BETA GAIN of a transistor that has an ALPHA GAIN of .99?

$$\text{BETA} = \frac{.99}{.01} = 99$$

Testing several transistors of the same type will result in a wide range in characteristics. However, the completion of this project will enable the reader to evaluate low-power transistors. Some units will be found to have low leakage and high beta values, whereas others will be just the reverse, according to the author's experience. By replacing the meter in the collector circuit with a pair of sensitive earphones, the reader may check the transistor for noise.

CURRENT GAIN ALPHA

Alpha is equal to the following ratio:

$$\frac{\text{Change in collector current}}{\text{Change in emitter current}}$$

and is expressed mathematically as follows:

$$\text{ALPHA} = \frac{\triangle I_c}{\triangle I_e}$$

Refer to Fig. 2-3. On closing the base circuit SW 2 observe the following reading: The collector meter reading (Ic) .4 ma. The emitter meter reading (Ie) .42 ma. Change in collector current = .4 ma - .02 ma = .38 ma. Change in emitter current = .42 ma - .02 ma = .4 ma.

$$\text{ALPHA} = \frac{.38}{.4} = .95$$

For junction transistors alpha is never greater than 1.

OHMMETER TEST

To check a transistor for FORWARD and REVERSE bias resistance, the ohmmeter leads are checked first for polarity, and then for voltage. On some ohmmeters the common meter terminal is positive; therefore, the red and black leads should be reversed to prevent confusion.

The voltage across the test leads when switched to any specific range should not exceed 7.5 volts. A check should be made using a voltmeter and selecting the range where the test lead potential is less than 7.5 volts. The following check was made using a Triplett meter 630NA; on the X1K range, the test potential measured 1.5 volts.

FORWARD BIAS

Collector to base—forward bias check (Fig. 2-4A):

Ohmmeter		Electrode
Red Lead (+)	to	Collector
Black Lead (−)	to	Base

The ohmmeter reading was 500 ohms, representing a relatively low forward resistance and recorded as satisfactory.

Emitter to base—forward bias check (Fig. 2-4A):

Ohmmeter		Electrode
Red Lead (+)	to	Emitter
Black Lead (−)	to	Base

The ohmmeter reading was 600 ohms, representing a relatively low forward resistance, and was recorded as satisfactory. (The forward bias applied to the collector in this test will not cause damage since the test is momentary and the test current small.)

REVERSE BIAS

Collector to base—reverse bias check (Fig. 2-4B)

Ohmmeter		Electrode
Red Lead (+)	to	Base
Black Lead (−)	to	Collector

Ohmmeter reading was a very slight deflection of the needle, representing a relatively high reverse resistance, and was recorded as satisfactory.

Emitter to base—reverse bias check (Fig. 2-4B)

Ohmmeter		Electrode
Red Lead (+)	to	Base
Black Lead (−)	to	Emitter

Ohmmeter reading was a very slight deflection of the needle, representing a relatively high reverse resistance, and was recorded as satisfactory. For NPN transistors reverse the ohmmeter leads and use the same test.

IN-CIRCUIT TRANSISTOR TEST

Let us assume that the amplifier illustrated in Fig. 2-5 was operating normally as a Class A amplifier, when suddenly the transistor developed an internal short or open circuit across

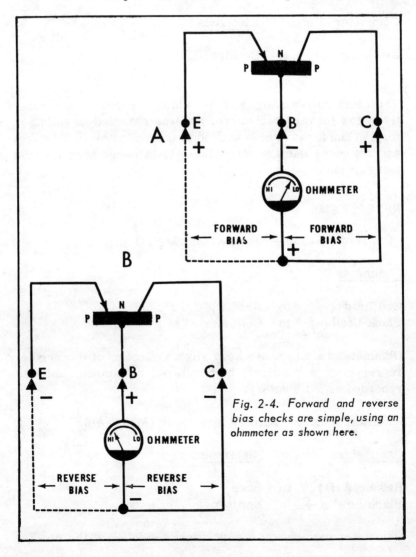

Fig. 2-4. Forward and reverse bias checks are simple, using an ohmmeter as shown here.

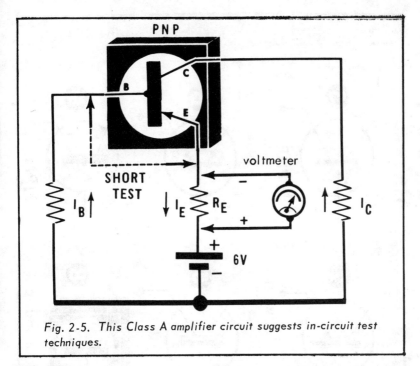

Fig. 2-5. *This Class A amplifier circuit suggests in-circuit test techniques.*

the emitter-base junction. In either case the collector current (I_c) would be cut off, with the exception of a small leakage current. Since I_c is approximately 95% of the emitter current, the voltage across the emitter resistor (R_E) would drop substantially.

This condition may readily be detected by a voltage measurement across the emitter resistor. For example, connecting a jumper wire between the base-emitter terminals will simulate an internal short-circuited base-emitter junction, thereby causing the voltage drop across R_E to decrease substantially. Upon removing the jumper wire the reading should return to normal. Should the emitter-base junction be open, the meter reading will indicate leakage current between the emitter and collector which should be in the order of a few microamperes. It is NOT advisable to make this test on power transistor stages!

TRANSISTOR SYMBOLS

The basic symbols of a triode vacuum tube and a three electrode transistor are shown for comparison in Fig. 2-6, dia-

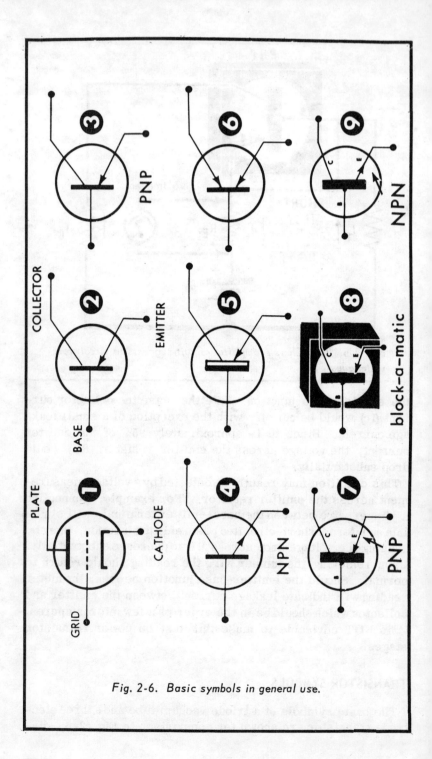

Fig. 2-6. Basic symbols in general use.

grams 1 and 2. In the transistor symbol the emitter electrode is represented by an arrow at an angle. The collector electrode is also drawn at an angle adjacent to the emitter. Both these designations intersect a line representing the base electrode.

If the emitter is a "P" type electrode, the arrow points to the base; if the emitter is "N" type, the arrow points away from the base. See diagrams 3 and 4. This standard method of identifying "P" type and "N" type emitters has not yet been universally adopted; however, it has been generally accepted that when the arrow points to the base it represents a "PNP" transistor; when pointing away from the base it represents an "NPN" transistor.

A minor variation of the basic transistor symbol is shown in diagram 5. Notice the slight change in the base representation. Diagram 6 is the symbol used to represent a symmetrical transistor, where the emitter and collector junctions have equal areas of contact. Such a device is used as a double diode in "AFC" circuits. The transistor for general applications has a collector area larger than that of the emitter. The transistor symbols shown in diagrams 7, 8, and 9 are used in this book to help further clarify circuit operation.

Chapter 3

Transistor and Vacuum Tube Equivalents

The electrodes of a transistor have operating functions similar to those of a vacuum tube. The order of their equivalence is given below and illustrated in Fig. 3-1.

VACUUM TUBE	TRANSISTOR
Cathode	Emitter
Control Grid	Base
Plate	Collector

The characteristic curves of both devices are similar. The family of curves in Fig. 3-2A represents the plate voltage-plate current characteristics of a typical pentode vacuum

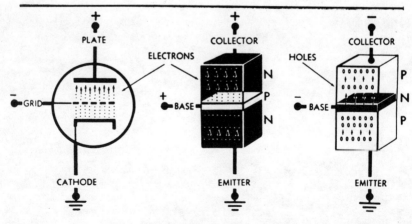

Fig. 3-1. These sketches illustrate the similarity of the operational functions of vacuum tubes and transistors.

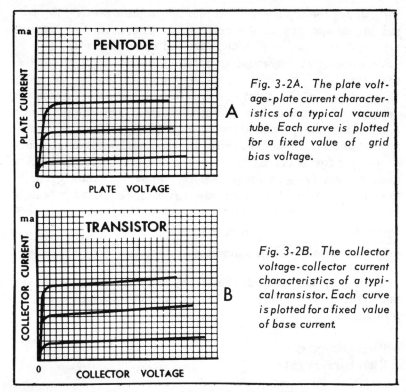

Fig. 3-2A. The plate voltage-plate current characteristics of a typical vacuum tube. Each curve is plotted for a fixed value of grid bias voltage.

Fig. 3-2B. The collector voltage-collector current characteristics of a typical transistor. Each curve is plotted for a fixed value of base current.

tube. A corresponding family of curves representing the collector voltage-collector current characteristics of a typical transistor is shown in Fig. 3-2B.

CIRCUIT EQUIVALENTS

There also are circuit similarities between vacuum tube and transistor stages; these circuit equivalents are listed as follows and illustrated in Fig. 3-3.

VACUUM TUBE	TRANSISTOR
Grounded Cathode	Grounded Emitter
Grounded Grid	Grounded Base
Grounded Plate	Grounded Collector

GROUNDED EMITTER

The most popular of the three circuits is the grounded emitter, where the input signal is applied to the base and the output signal appears at the collector. The grounded element (emitter) is common to both input and output circuits.

Some confusion may exist as to which electrode of a transistor should be grounded. This confusion is due to the term "grounded," which in this case refers to the point of common return of both input and output circuits. Any one point of a circuit may be grounded to the chassis regardless of circuit arrangement. Ground is used by the troubleshooter as a reference point for checking input and output circuits. It also is used for specifying a voltage as being "above or below ground"; therefore, it provides a convenient reference point for testing both vacuum tube and transistor circuits. The term "grounded" generally is used interchangeably with "common."

GROUNDED OR COMMON EMITTER

In this amplifier circuit the signal is applied to the base and taken from the collector. The gain characteristics are as follows:

- High voltage gain
- High current gain

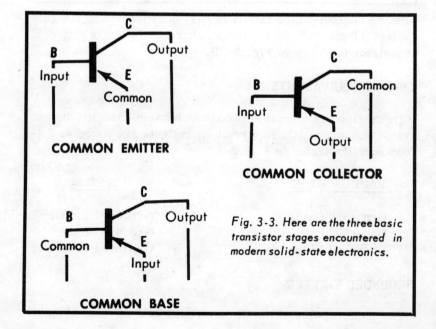

Fig. 3-3. Here are the three basic transistor stages encountered in modern solid-state electronics.

- Highest power gain
- Input resistance: 600 ohms
- Output resistance: 20,000 ohms

GROUNDED OR COMMON BASE

Here, the signal is applied to the emitter and taken from the collector. The gain characteristics are as follows:

- High voltage gain
- Current gain less than unity
- Medium power gain
- Input resistance: 60 ohms
- Output resistance: 100,000 ohms

GROUNDED OR COMMON COLLECTOR

In this circuit the signal is applied to the base and taken from the emitter. The gain characteristics are as follows:

- Voltage gain less than unity
- High current gain
- Lowest power gain
- Input resistance: .5 megohm
- Output resistance: 20,000 ohms

GROUNDED EMITTER

The grounded-cathode and grounded-emitter circuits are shown for comparison in Fig. 3-4. The difference in operation between these two stages is as follows:

VACUUM TUBE: Any small change in grid bias voltage will be accompanied by a substantial change in plate current.

TRANSISTOR: Any small change in base bias current will be accompanied by a substantial change in collector current. These two basic points are illustrated in Fig. 3-5. In the vacuum tube grid circuit a potentiometer is used to illustrate grid voltage control; in the transistor base circuit a rheostat is used to illustrate base current control. The applied bias in both cases is usually fixed to provide an operating point for the input signal.

The grounded-emitter amplifier, illustrated in Fig. 3-4, shows the base-emitter junction biased in the forward direction. The negative potential at the "N" type base repels the free electrons towards the base-emitter junction, overcomes the potential barrier and establishes a fixed base current, the magnitude of which is governed by the resistance of R2. The collector is negative with respect to the emitter; therefore, the positive charges in the "P" type collector are attracted away from the base-collector junction, thereby establishing

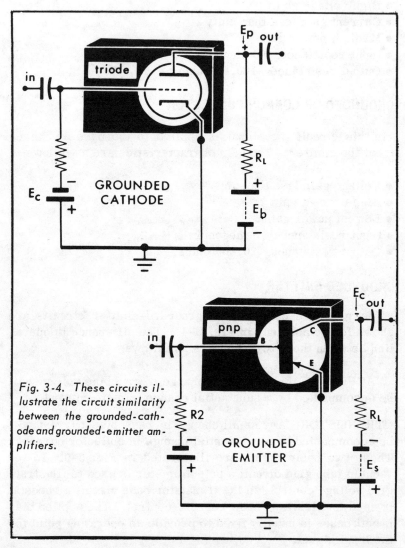

Fig. 3-4. These circuits illustrate the circuit similarity between the grounded-cathode and grounded-emitter amplifiers.

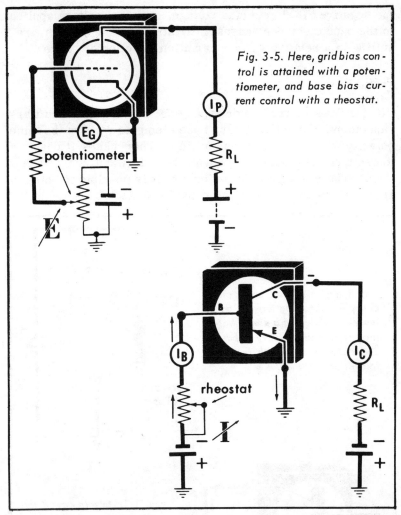

Fig. 3-5. Here, grid bias control is attained with a potentiometer, and base bias current control with a rheostat.

a reverse bias effect. Notice that both base and collector electrodes have the same polarity with respect to the emitter.

When the forward bias is active, approximately 98% of the emitter current will pass through the base electrode to the collector circuit; therefore, the magnitude of the collector current will be governed by the magnitude of the forward bias. The base current is approximately 2% of the emitter current for the above condition.

The "DC" base current (bias) is usually fixed to establish a "DC" operating point for the input signal. In the case of the vacuum tube amplifier, the operating point is established by

41

the negative "DC" grid bias voltage. A signal voltage applied to the base electrode causes a variation in base current, resulting in a relatively large variation in collector current.

FIXED BIAS

Regardless of transistor type (PNP or NPN) or circuit arrangement, the collector and base elements require the same polarity with respect to the emitter. This makes it possible to use a single battery to provide both forward and reverse bias. The reader may question how it is possible to obtain a reverse bias when the collector and base electrodes are of the

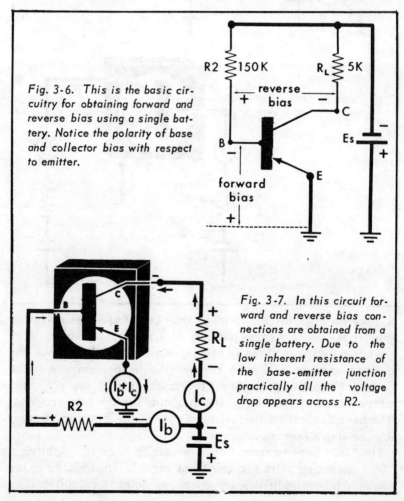

Fig. 3-6. This is the basic circuitry for obtaining forward and reverse bias using a single battery. Notice the polarity of base and collector bias with respect to emitter.

Fig. 3-7. In this circuit forward and reverse bias connections are obtained from a single battery. Due to the low inherent resistance of the base-emitter junction practically all the voltage drop appears across R2.

same polarity. This is explained by the fact that the collector voltage is greater than the base voltage with respect to the emitter; therefore, a difference of potential will exist between the base and collector electrodes, representing the reverse bias. See Fig. 3-6.

FORWARD BIAS (SINGLE BATTERY)

In the case of the PNP transistor circuit shown in Fig. 3-7, resistor R2 is connected between the negative terminal of the battery and the base. The emitter being positive will establish a base current, the amount of which is determined by the resistance of R2. This takes care of the forward bias requirement and provides a fixed operating point for the input signal.

Measuring the voltage drop across R2 provides a convenient method of evaluating the base current. The inherent resistance of the base-emitter junction is very much smaller than R2 and may be disregarded in the base current calculation. The forward-bias voltage may be evaluated by subtracting the measured voltage drop across R2 from the battery voltage (Es). This voltage is usually a few tenths of a volt due to the relatively low inherent resistance of the junction.

PHASE INVERSION

For PNP transistors the collector voltage is negative with respect to ground, and any decrease in this voltage due to signal input is referred to as positive-going. When a signal voltage is applied to the base of a grounded-emitter stage, the signal appears in the collector circuit, where it is amplified and phase-inverted.

For example, when a positive-going signal is applied to the base of a PNP stage, it opposes the negative forward bias, resulting in a decreasing base current. A decreasing base current, in turn, causes a decreasing collector current and a smaller voltage drop across the load resistance R_L. This produces a negative-going signal at the collector.

SUMMARY

The base electrode may be considered as the "nerve" of the grounded-emitter amplifier. A slight variation of the base current releases a collector current many times its value from the battery. The released energy appears in the collector circuit as current variations that may be transformer-coupled or resistance-capacity-coupled to a following stage for further amplification. The grounded-emitter stage features amplification and phase inversion; this corresponds closely to the grounded-cathode amplifier.

EFFECT OF EMITTER RESISTOR ON FORWARD BIAS

Returning to Fig. 2-1, remove or short-circuit the protec-

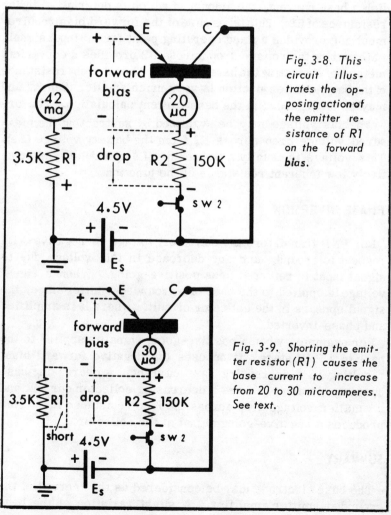

Fig. 3-8. This circuit illustrates the opposing action of the emitter resistance of R1 on the forward bias.

Fig. 3-9. Shorting the emitter resistor (R1) causes the base current to increase from 20 to 30 microamperes. See text.

tive resistor R1 and calculate the base current. See Fig. 3-8. In the experiment a reading of 20 microamperes was recorded; therefore, shorting R1 has caused the base current to increase to 30 microamperes. Compare Fig. 3-9 with Fig. 3-8.

Notice that R1 is 3.5K and is in series in the base current loop with R2 which is 150K. It would seem, since R1 is a relatively low resistance, that its removal would have very little effect on the base current. However, in the test when the base circuit was closed, the emitter current through R1 was .42 ma. This develops a relatively large voltage drop across R1 in opposition to the battery voltage Es, reducing the base current appreciably. Although R1 was used for meter protection in the experiment, it has another important use in the grounded-emitter circuit yet to be discussed.

COLLECTOR LOAD RESISTOR

It has been stated that a small change in base current will be accompanied by a relatively large change in collector current. Therefore, if a small AC potential is applied across the base-emitter junction, it will have an appreciable effect on the base current. This in turn will produce a relatively large AC component in the collector current. It is possible to convert this AC current component to an AC voltage by connecting a load resistor (R_L) in series with the collector electrode. See Fig. 3-10. This provides a convenient method of capacitively-coupling the output to the input of the following stage, which could be another transistor or a vacuum tube; thus, transistors are compatible with vacuum tubes in hybrid electronic equipment.

DISTORTION IN TRANSISTOR AMPLIFIERS

Transistors, like vacuum tubes, have inherent distortion characteristics. Referring to Fig. 3-11, it will be seen that a pure sine-wave signal is applied to the base of a grounded-emitter amplifier. An amplified and phase-inverted version of this signal appears at the collector, as shown in Oscillogram 2. Notice the slight distortion of the positive peaks as compared with the negative peaks. This is due to the nonlinear rise in collector current for a linear rise in base current, which is inherent in transistors at relatively high values of base current.

This is further illustrated in Oscillogram 3. Here, the input signal level is increased slightly to show a pronounced distortion of the positive peaks, while the negative peaks remain normal. A further increase of input signal causes clipping (flat tops) of the positive peaks. See Oscillogram 4. This is called saturation, a point reached when the collector voltage approaches zero; therefore, despite an increase in base current, no further increase in collector current takes place.

Further increasing the input signal causes clipping of the negative peaks, also. This condition is called "cut-off" and is reached when the positive swing of the input signal equals the fixed negative base bias, causing it to drop to zero. In some cases the input signal drives the base positive, a condition referred to as "beyond cut-off" for PNP transistors. In NPN transistors, the polarity causing saturation and cut-off are reversed, of course.

A study of Oscillogram 5 shows that clipping of both negative and positive peaks appears almost simultaneously as the input signal is increased. This indicates that the fixed bias is almost centered between cut-off and saturation. The collector signal appears as a true amplified reproduction of the input signal by reducing the input signal to a normal level. The gain

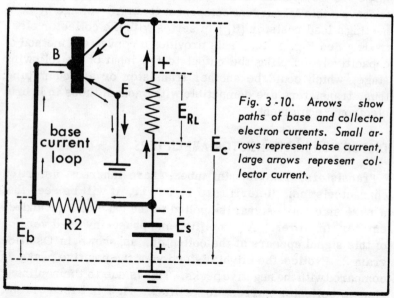

Fig. 3-10. Arrows show paths of base and collector electron currents. Small arrows represent base current, large arrows represent collector current.

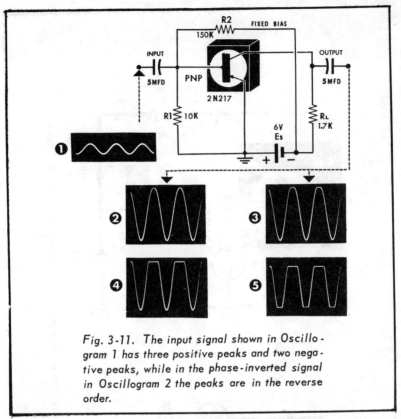

Fig. 3-11. The input signal shown in Oscillogram 1 has three positive peaks and two negative peaks, while in the phase-inverted signal
in Oscillogram 2 the peaks are in the reverse
order.

of the transistor tested was about 40; therefore, the collector
voltage amplitude was attenuated 10 to 1 and would be 10 times
greater than shown in the Oscillograms.

Fig. 3-12 demonstrates the voltage amplification and phase
reversal characteristics of a grounded-emitter amplifier.
Toggle switch SW2 is used to open the base circuit to permit
measurement of the leakage current (Iceo)*. This check is
necessary before continuing with the study. Transistors with
excessive leakage should be discarded.

The 100K potentiometer is connected in the base circuit to
permit adjustment of the base current. The 50K resistor in
series with the potentiometer provides a minimum base circuit resistance. The 1K resistor in the collector circuit provides a load resistance.

---

*Iceo is the amount flowing between the emitter and collector with the
base open. See Fig. 3-11.

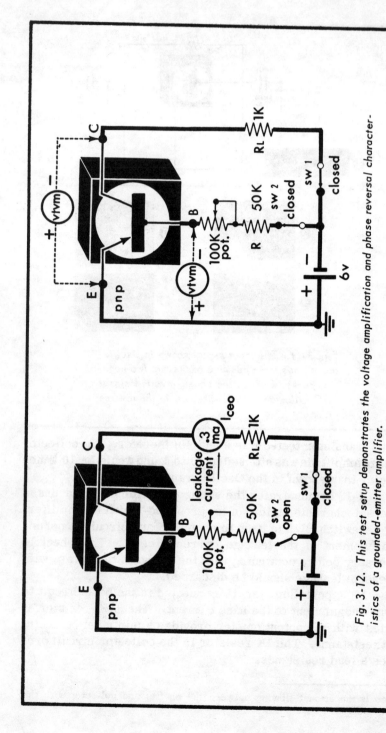

Fig. 3-12. This test setup demonstrates the voltage amplification and phase reversal characteristics of a grounded-emitter amplifier.

Checking Collector and Base Voltages

1. Close SW1 and open SW2 and record leakage current. Then close SW2.
2. Connect a VTVM (3-volt range) between emitter and base, positive test lead to emitter. See Fig. 3-12.
3. Rotate the base-current potentiometer through its range and record minimum and maximum input voltage. These readings will be less than .5 volt
4. Connect the VTVM (10-volt range) between collector and emitter, negative test lead to collector.
5. Rotate the base-current potentiometer through its range and record minimum and maximum output voltage.
6. Compare the change of output voltage to change of input voltage and observe voltage gain. Notice that when the base voltage increases negatively, the collector voltage becomes less negative, indicating a phase reversal.

Chapter 4

Transistor Biasing Techniques

An increase in the temperature of semiconductor material causes a breakdown of some covalent bonds. This disruption of the crystal lattice structure releases electrons which act as current carriers, thereby increasing the conductivity of the crystal. As the temperature increases, the presence of the additional carriers results in an approach to intrinsic conductivity. This means that the conductance of the crystal depends more upon the properties of the germanium and less upon its impurities, making it useless for transistor operation. Transistor characteristics depend upon the normal conductivity of semiconductor material; therefore, an increase in temperature is undesirable.

Temperature stability in transistors is of utmost importance, since a rise in temperature increases the leakage current (Ico). In grounded-emitter circuits this current flows through the base electrode and is amplified by the beta factor. For example, if the beta amplification factor is 20, an increase in Ico will be amplified 20 times, causing an increase in collector current and a further increase in temperature which can eventually destroy the transistor. To maintain stability some means of self-bias control must be established. This can be accomplished by using the "DC" negative feedback principle.

An increase in temperature also may cause a reduction of a transistor's input "DC" resistance, resulting in an increase of base current. This is of particular concern in transformer-coupled power transistors where the inherent base resistance constitutes the major portion of the input circuit. Any appreciable reduction of this resistance would cause a "runaway" condition and damage the transistor.

NEGATIVE FEEDBACK

Negative feedback occurs when a portion of the output current is fed back in opposite phase or polarity to the input circuit. Fixed bias supplies a constant operating bias (voltage or current) uninterrupted by signal amplitudes or changes in "DC" collector voltage, but it does not provide any form of feedback. Self bias supplies operating bias and "DC" negative feedback when the bias resistor is bypassed. If not bypassed, it will supply both "AC" and "DC" negative feedback. "AC" negative feedback is used primarily to improve frequency response and reduce distortion. "DC" negative feedback is used also to prevent excessive changes in the "DC" operating point of a transistor stage.

FIXED BIAS

Fig. 4-1 shows a grounded-emitter circuit using a fixed-bias resistor (R2) connected from the negative terminal of the battery to the base electrode. This bias method is the least satisfactory since there is a variation in characteristics between comparable transistors. Due to these irregularities the bias resistor (R2) would require adjustment for each stage; even then the bias would shift with changes in temperature. However, with low collector voltage and current this method would operate normally at room temperature.

EMITTER BIAS

Fig. 4-2 shows the same circuit using "DC" negative feedback; notice the addition of R1 in the emitter lead. In the grounded-emitter circuit the emitter is common to both the input and output circuits; therefore, a resistance inserted in the emitter lead will provide a stabilizing influence on the collector current. Also:

1. The base electrode is negative with respect to the emitter. For "PNP" transistors this constitutes a forward bias.

2. The current flowing through the emitter resistor (R1) develops a negative voltage at the emitter which opposes the forward bias applied to the base.

All of the collector current flows through the emitter resistor (R1); therefore, any excessive shift in this current is opposed by the emitter bias it develops. This action is "DC" negative feedback and is similar to cathode bias of a vacuum tube. To avoid "AC" negative feedback the emitter resistor (R1) should be bypassed to prevent a loss in gain. "AC" negative feedback may be desired in some cases to improve frequency response and minimize distortion. However,

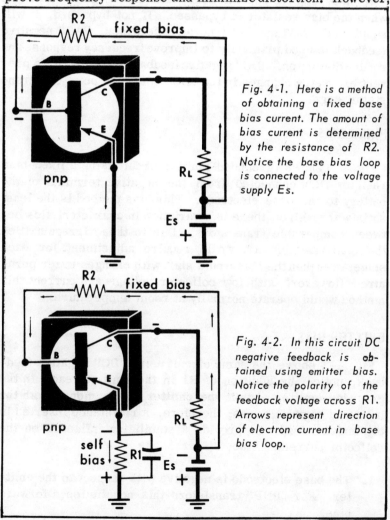

Fig. 4-1. Here is a method of obtaining a fixed base bias current. The amount of bias current is determined by the resistance of R2. Notice the base bias loop is connected to the voltage supply Es.

Fig. 4-2. In this circuit DC negative feedback is obtained using emitter bias. Notice the polarity of the feedback voltage across R1. Arrows represent direction of electron current in base bias loop.

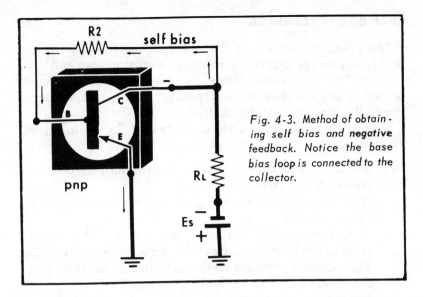

Fig. 4-3. Method of obtaining self bias and *negative* feedback. Notice the base bias loop is connected to the collector.

since we are interested at the moment in maintaining a constant "DC" operating point, R1 is shown bypassed in Fig. 4-2.

COLLECTOR-TO-BASE BIAS LOOP

Another method of "DC" negative feedback is shown in Fig. 4-3. In this circuit the base bias resistor (R2) is connected directly to the collector. In this position the resistor performs two operations; it determines the "DC" base bias current and provides "DC" negative feedback.

The stabilization action of this circuit may be described as follows:

1. A small increase in leakage current causes a relatively large increase in collector current.
2. This causes the voltage across the load resistor (R_L) to increase and the collector voltage (Ec) to decrease.
3. The base current, which is dependent upon the collector voltage, also decreases and opposes the rise in collector current.

Besides compensating for the effect of temperature, "DC" negative feedback also compensates for the difference in characteristics between comparable type transistors.

SELF BIAS vs FEEDBACK

The primary function of the self-bias resistor (R2) in Fig. 4-3 is to provide negative feedback, so that a fixed collector current is established. Since this resistance (R2) is determined by the specified base current it is fairly large; therefore, the feedback will be limited.

A circuit for providing stricter control of the base current is shown in Fig. 4-4, and may be described as follows:

1. To increase negative feedback control the self-bias resistor (R2) is reduced in value.
2. This tends to increase the base bias current beyond its fixed operating point.
3. To compensate for this increase, a reverse bias is applied to the base by battery Eb through resistance R3. This network provides a more effective DC control, since the forward bias resistor R2 is reduced.

The disadvantage of this type of circuit is the need for an additional battery. However, by changing the relative positions of the load resistor (R$_L$) and the supply battery (Es), a forward and reverse base bias may be obtained from a single battery, as shown in Fig. 4-5.

TRANSISTOR CHARACTERISTIC CURVES

The collector voltage-collector current characteristics for

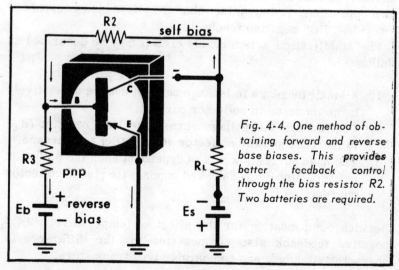

Fig. 4-4. One method of obtaining forward and reverse base biases. This provides better feedback control through the bias resistor R2. Two batteries are required.

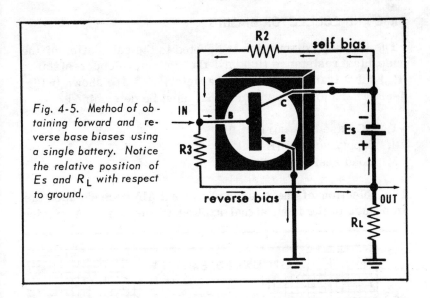

Fig. 4-5. Method of obtaining forward and reverse base biases using a single battery. Notice the relative position of Es and R_L with respect to ground.

different values of base bias current are shown in Fig. 4-6. These curves resemble those of a pentode vacuum tube, the difference being that the transistor curves are shown for constant base bias currents, while the vacuum tube curves are based on constant grid bias voltages. The values along the voltage and current axes are shown with negative signs. These signs indicate that the curves are for a PNP transistor, since its collector voltage is negative with respect to the emitter. For NPN transistors the voltage and current signs are positive.

In a PNP transistor the emitted current is composed of "holes" (positive carriers); therefore, the direction of current through the external circuit is from the emitter connection to the collector. Electrons (negative carriers) comprise the emitted current in an NPN transistor; therefore, the direction of current through the external circuit is from the collector connection to the emitter, similar to the current flow through a vacuum tube.

Conventional current is considered to flow from the positive terminal to the negative terminal of the load resistor, similar to hole conduction, while electron current flow is in the opposite direction. Those familiar with Kirchoff's Law know that the direction of current is a matter of choice. However, in the study of vacuum tube and transistor theory, the direction of electron current flow is popular among technicians.

BASE AND COLLECTOR BIASING

The following discussion is devoted to the calculation of the base bias resistance (R2) and the collector load resistance (R_L) for a transistor whose characteristics are shown in Fig. 4-6. There are three values that must be determined:

1. DC base bias current (Ib)
2. Supply voltage (Es)
3. Load resistance (R_L)

The selection of the base bias current (Ib) depends upon the amplitude of the input signal applied to the base. Assuming

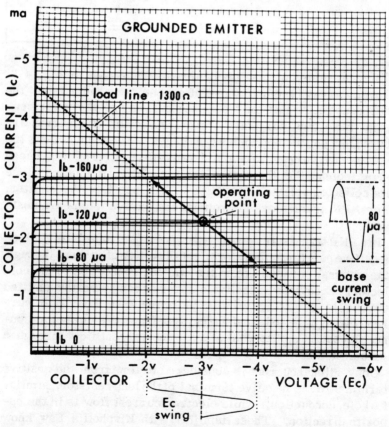

Fig. 4-6. Collector voltage-collector current characteristics of a transistor selected for discussion. Notice the base current swing due to a small input signal and the corresponding swing of collector voltage.

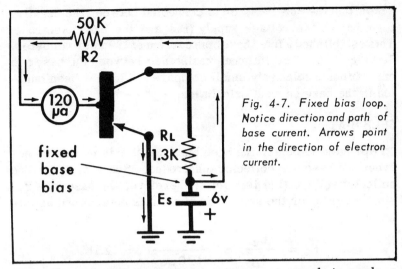

a small input signal of 80 microamperes, peak to peak, a fixed base current of 120 microamperes has been selected for the DC operating point. See Fig. 4-6.

The voltage supply (Es) is fixed at 6 volts and the collector potential (Ec) has been set at -3 volts (PNP), midpoint on the collector voltage axis. A line drawn from -6 volts on the collector voltage axis, through the operating point, and then extended to the collector current axis represents the load line. The arrows pivoted on the operating point in Fig. 4-6 represent the portion of the load line required by the small signal input of 80 microamperes p-p. Notice that where the load line passes through the operating point (120 microamperes), the collector potential is -3 volts and the collector current is 2.28 milliamperes; therefore, the load resistance is determined as follows:

$$R_L = \frac{E_s - E_c}{I_c} = \frac{3v}{2.28ma} = 1.3K$$

FIXED BIAS

For the fixed-bias connection shown in Fig. 4-7, the base bias resistance (R2) is determined as follows:

$$R_2 = \frac{E_s}{I_b} = \frac{6v}{120\mu a} = 50K$$

Notice that the 50K resistor is connected between the negative terminal of the voltage supply (Es) and the base electrode. This established a fixed base bias current of 120 microamperes. See Fig. 4-6. The internal resistance between the base and emitter being relatively small compared to R2 has been omitted in the base current calculation.

SELF BIAS

To provide self bias, the base bias resistor is connected between the base and collector electrodes. See Fig. 4-8. The collector voltage (Ec) is 3 volts; therefore, the base bias resistance (R2) for the same bias current is determined as follows:

$$R_2 = \frac{E_c}{I_b} = \frac{3v}{120\mu a} = 25K$$

Notice the resistance of R2 is decreased from its fixed bias values of 50K to 25K.

SELF BIAS PLUS REVERSE BIAS

A circuit that provides stricter feedback control of the DC operating level is shown in Fig. 4-9. Notice that the base resistor (R2) has been reduced from 25K to 15K to provide greater DC negative feedback. However, reducing the resistance to 15K will cause the current through R2 to increase while still maintaining a base bias current of 120 microamperes.

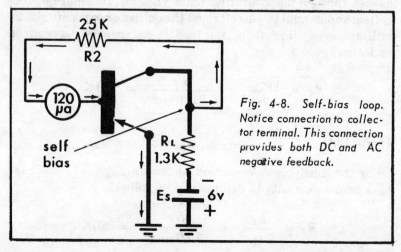

Fig. 4-8. Self-bias loop. Notice connection to collector terminal. This connection provides both DC and AC negative feedback.

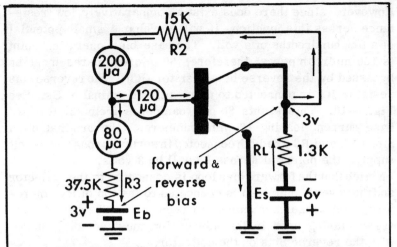

Fig. 4-9. This circuit offers improved negative feedback control. Notice the increase of bias loop current and the shunting effect of the reverse bias circuit.

Maintaining the normal level of the base current is accomplished by the reverse bias of battery Eb and resistor R3. See Fig. 4-9. The amount of reverse bias current required is 200 - 120 = 80 microamperes which is determined by the resistance of R3; therefore:

$$R_3 = \frac{E_b}{I_{R_3}} = \frac{3v}{80 \mu a} = 37.5 K$$

SINGLE BATTERY BIAS SYSTEM

A system using a single battery to provide both forward and reverse base bias is shown in Fig. 4-10. By changing the position of R_L and Es with respect to ground, the forward and reverse bias network of R2 and R3 operates as follows: the potential applied to the forward bias resistor R2 is -3 volts; therefore, 15K ohms at 3 volts = 200 microamperes:

$$I_f = \frac{E_c}{R_2} = \frac{3v}{15K} = 200 \mu a$$

f — forward bias

The resistance of the emitter-base junction is relatively small and has been disregarded in the base bias calculation.

However, since there does exist a comparatively low resistance across this junction, it will develop a small potential of a few hundredths of a volt. The base bias operating point is 120 microamperes; therefore, 80 microamperes must be bypassed by the reverse bias resistor (R3). The reverse bias resistor R3 is connected to the positive terminal of Es. See Fig. 4-10. This shunts 80 microamperes from the forward base current leaving 120 microamperes, the original bias. Since R2 and R3 are both connected in series across the 6-volt supply, the potential across R3 will be 3 volts.

Notice that the forward bias loop is connected to the collector and the reverse bias loop is connected to the top of R_L; therefore, both biases are subject to negative feedback. The reverse biasing used in this discussion must not be confused with the reverse bias on the collector.

Fig. 4-11 shows the base bias current supplied from a tap on voltage divider R2 and R3. This network, in conjunction with emitter stabilizing resistor R1, provides good stability. Fig. 4-12 shows voltage divider R2 and R3 connected between the collector and ground to provide self bias. Notice the emitter is connected directly to ground.

PRECAUTIONS

Sudden voltage surges can damage transistors. When replacing a transistor it is advisable to remove the applied voltage.

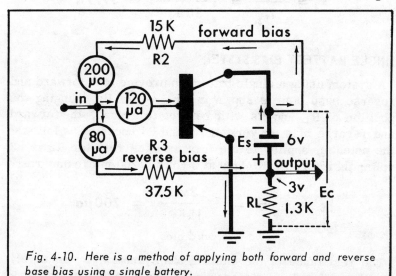

Fig. 4-10. Here is a method of applying both forward and reverse base bias using a single battery.

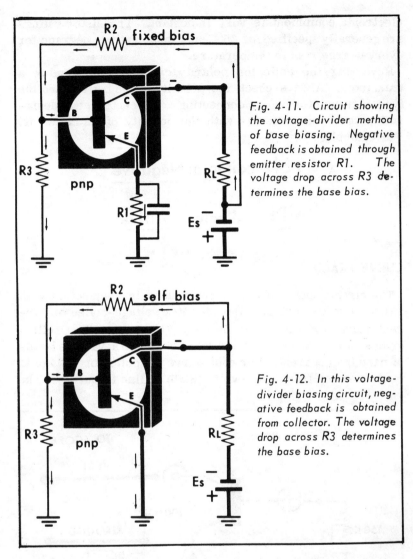

Fig. 4-11. Circuit showing the voltage-divider method of base biasing. Negative feedback is obtained through emitter resistor R1. The voltage drop across R3 determines the base bias.

Fig. 4-12. In this voltage-divider biasing circuit, negative feedback is obtained from collector. The voltage drop across R3 determines the base bias.

This is done by a switch or disconnecting the battery. Remove the transistor before measuring circuit resistances. When using an ohmmeter to measure a transistor be sure to determine the polarity of the internal battery of the meter.

Since high temperatures can damage transistors, when soldering transistor leads it is a good idea to hold the lead being soldered with a pair of pliers to provide a suitable heat sink. Transistors behave erratically when placed near circuit components that generate heat; therefore, their location in electronic equipment is very important. Transistor ratings are generally specified for 25°C and these ratings degrade for every degree rise in temperature.

Reversing the collector polarity can seriously damage a transistor. Always check the polarity of the battery and the type of transistor before connecting. The center letter designating the base conforms with the polarity of its collector. For example:

P N P collector Negative

N P N collector Positive

CURVE TRACING

The circuit shown in Fig. 4-13 is used in conjunction with an oscilloscope to display the collector voltage-collector current characteristic curves shown in Fig. 4-14. The Oscillograms in Fig. 4-14 are the patterns observed at various levels of base bias current. For each curve the collector voltage is made to sweep from 0 to 6 volts (RMS) at line frequency. The

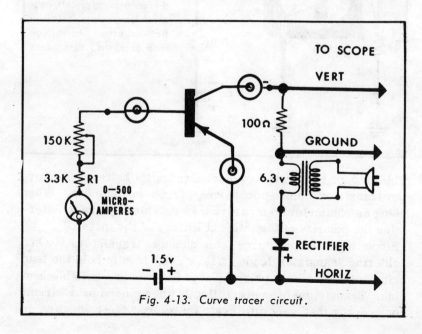

Fig. 4-13. Curve tracer circuit.

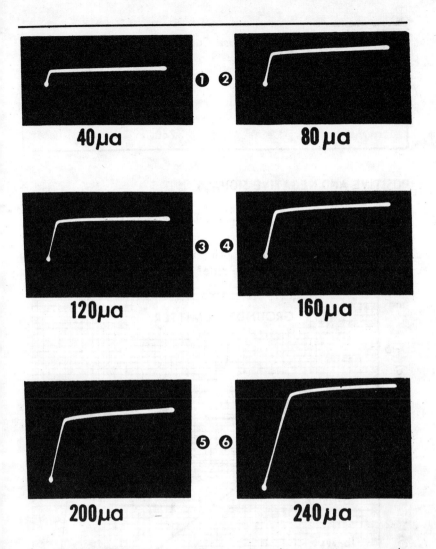

Fig. 4-14. Collector voltage-collector current characteristics curves obtained with an oscilloscope and the circuit in Fig. 4-13.

horizontal sweep represents the collector voltage axis, and the vertical sweep represents the collector current axis for one sweep period. Plotting several curves for different values of base current provides a family of curves as shown in Fig. 4-15.

The Oscillograms shown in Fig. 4-14 are for each 40 microampere increase in base current and are recorded as follows:

Oscillogram	Base Current (microamps)
1	40
2	80
3	120
4	160
5	200
6	240

POSITIVE AND NEGATIVE SIGNALS

Terms such as "positive-going" and "negative-going" must not be confused with the positive and negative half sine waves, since each half sine wave is both positive- and negative-going. For example, as a sine wave rises from zero towards a posi-

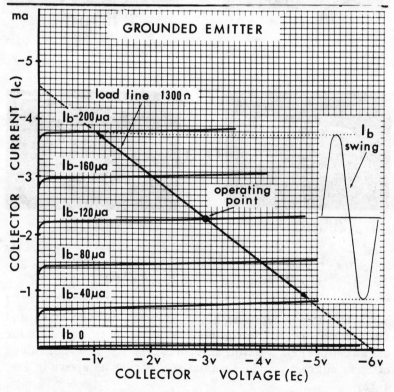

Fig. 4-15. Collector voltage-collector current characteristics curves for six values of base current.

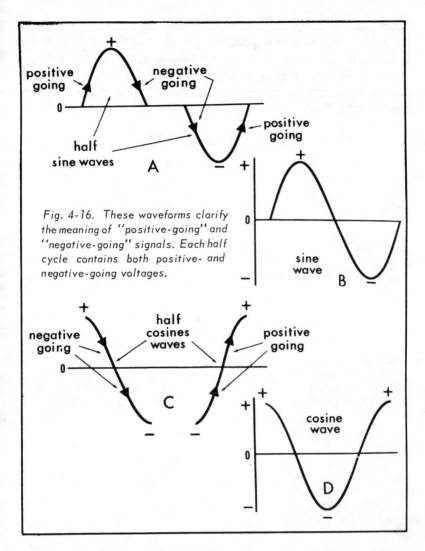

Fig. 4-16. These waveforms clarify the meaning of "positive-going" and "negative-going" signals. Each half cycle contains both positive- and negative-going voltages.

tive peak it is "positive-going," and when falling from a positive peak it is "negative-going." When rising from zero towards a negative peak it is "negative-going," and when falling from a negative peak to zero it is "positive-going." See Fig. 4-16A and B. Therefore, a sine wave excursion from a positive peak to a negative peak (a half cosine wave) is "negative-going," and the excursion from a negative peak to a positive peak is "positive-going."

The above mentioned signal swings also apply to square waves, and all wave motions containing DC components. For example, a potential that rises from +50 volts to +100 volts is going more positive (positive-going), and when falling from +100 volts to +50 volts is going less positive or "negative-going." When a signal containing a DC component (average value) is coupled by a capacitor, the DC or average value is eliminated.

Chapter 5

Transistor Amplifiers

The characteristics of the three transistor amplifier circuits discussed in Chapter 3 indicate that the typical grounded-emitter stage meets all the gain requirements for voltage, current, and power amplification. Two or more of these stages may be coupled in cascade either by capacitor or transformer, and in some instances PNP type transistors are coupled directly to NPN types or vice versa. The low input and relatively high output impedance of the grounded-emitter stage is just the reverse of its vacuum tube equivalent, the grounded-cathode circuit. Fig. 5-1 shows a grounded-emitter stage using the DC biasing techniques described in the previous Chapter. Fig. 5-2 shows two RC-coupled grounded-emitter stages terminated with a pair of earphones.

CAPACITIVE COUPLING

The input impedance of a grounded-emitter stage is approximately 600 ohms for a typical transistor; therefore, the coupling capacitor must be large enough to pass the lowest frequency of the applied signal. When the reactance of a coupling capacitor is equal to the input impedance, the signal drops 3 db, corresponding to a 30% loss in signal amplitude.

To obtain satisfactory coupling the reactance of C1 (Fig. 5-3) at the lowest frequency must be much less than the input impedance. For example, if the lowest frequency is 100 Hz and the input impedance is 1,000 ohms, the coupling capacitor should be at least 5 mfd. Since the voltages involved are small the DC working potential of this capacitor can be as low as 3 volts.

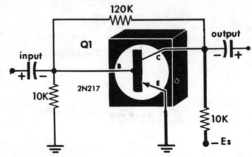

Fig. 5-1. A grounded-emitter stage biased for low-level Class A signal input.

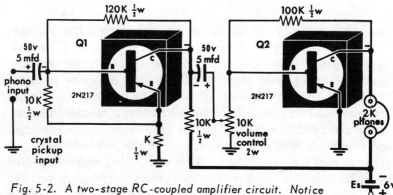

Fig. 5-2. A two-stage RC-coupled amplifier circuit. Notice that transistor Q2 is biased for a higher level input signal.

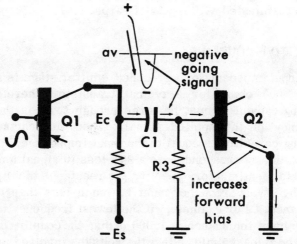

Fig. 5-3 Signal path through the coupling circuit between Q1 and Q2 for a negative-going signal.

In vacuum tube theory we learned that the coupling capacitor conveyed the signal <u>voltage</u> to the grid of the following stage. Accordingly, you may question why a capacitor is used to drive a current-operated device. The fact is that although a capacitor is voltage-operated, it does involve charge and discharge currents, which can be conveyed to the input circuit of the following stage. The charge and discharge of the coupling capacitor shown in Fig. 5-3 occurs in this manner:

1. Let us assume that without a signal the collector potential (Ec) of Q1 is -3 volts (PNP). In this case C1 will be charged initially to 3 volts.
2. When the collector current of Q1 decreases, the collector voltage increases negatively (PNP) and the charge on C1 increases.
3. The charge current flows through the base-emitter junction and aids the forward bias of Q2. This constitutes a negative-going signal.
4. When the collector current of Q1 increases, the collector voltage decreases, and discharges C1.
5. The discharge current flows through the base resistor (R3) and opposes the forward bias of Q2. This constitutes a positive-going signal. See Fig. 5-4.

As with vacuum tubes, the function of the coupling capacitor is to convey the AC current component and block the DC current component at the collector of Q1.

VOLUME CONTROL

A manual volume control is necessary in audio amplifiers to provide a means for adjusting the sound level. The control may be connected in the input or output circuit of any stage. However, in audio systems using several stages in cascade, it is good practice to connect the volume control in the input or output circuit of the first stage, because a control in this low-level position will help prevent overloading of the high-level stages due to strong input signals.

A volume control usually operates in conjunction with a coupling capacitor. The purpose of such a capacitor is to isolate or block the bias DC, thereby preventing interference with the DC operating point. The output of Q1 in Fig. 5-5A is coupled to Q2 through C1 and a volume control. The circuit connections of this control are very important, since its adjustment must not interfere with the DC bias constants. Fig. 5-5B clearly illustrates this point. At A, when the coupling capacitor is connected to the moving arm of the potentiometer, the DC base current of Q2 remains constant during adjustment of the control. However, at B when the moving arm of the control is connected to the base electrode, adjustment of the control will cause the DC base and collector currents of Q2 to shift simultaneously. This condition must be avoided.

INTERSTAGE MATCHING

The high output impedance of Q1 (Fig. 5-6) coupled to the low input impedance of Q2 presents a mismatch problem, resulting in a loss of gain. This can be overcome by using a stepdown transformer for impedance matching as shown in Fig. 5-6. However, in comparing the two types of coupling, the RC method offers better frequency response and is far more economical. In fact, it is often less expensive to add an extra RC stage to compensate for the loss in gain.

GROUNDED COLLECTOR

The output of a two-stage RC-coupled voltage amplifier may

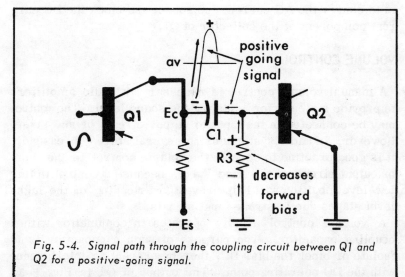

Fig. 5-4. Signal path through the coupling circuit between Q1 and Q2 for a positive-going signal.

be directly coupled to a power output stage by using a grounded-collector stage. Such a circuit is shown in Fig. 5-7. Since this stage resembles the cathode follower, it is generally referred to as an emitter-follower.

You'll recall that the grounded-collector stage has less than unity voltage gain, low power gain, but high current gain. Its input impedance is high and its output impedance low. The input impedance base-to-collector is high due to the reverse bias applied between these two electrodes. The output impedance is very low, generally less than 100 ohms, and like its vacuum tube equivalent it does not invert the phase of the input signal.

A grounded-collector stage also has low distortion due to its large negative feedback. It may be driven by an RC-coupled driver, and its high current gain and low output impedance permit direct coupling to the input of a power amplifier. Fig. 5-8 shows the grounded-collector stage RC-coupled to the output of the two-stage amplifier shown in Fig. 5-6. In Fig. 5-9 the grounded-collector is directly coupled to a power output stage. In this case, the emitter of the follower stage supplies the base signal current directly to the power stage.

The forward and reverse DC biasing requirements for the grounded-collector are the same as those required for the grounded-emitter. However, since the follower stage does not invert the phase of the signal it cannot supply its own shunt negative feedback. A further study of the circuit in Fig. 5-9 shows that a phase-inverted signal appears at the collector of the power output stage; therefore, a feedback loop from the collector of this stage to the base of the follower will satisfactorily meet the base bias and negative-feedback requirements.

TRANSISTOR POWER RATINGS

So far we have discussed only current and voltage amplification, but like vacuum tubes transistors can be made to operate as power amplifiers. The maximum power output of a transistor is rated at a specified operating temperature, generally $25^{\circ}C$, and is then derated for each degree rise in temperature. For example, in the 2N217 transistor data sheet the maximum power rating is 150 milliwatts at $25^{\circ}C$. This maximum power specification is then derated 3 milliwatts for

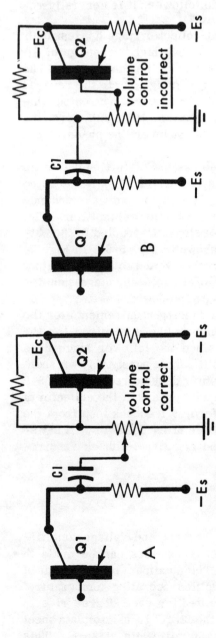

Fig. 5-5. A: CORRECT method of connecting a volume control. Notice the constant resistance of bias network regardless of the control arm position. B. INCORRECT volume control connection.

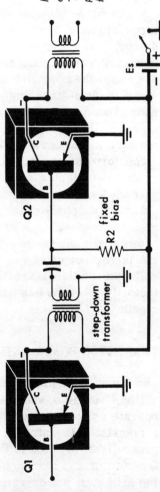

Fig. 5-6. A two-stage transformer-coupled transistor amplifier. This coupling system provides precise matching but lacks in frequency response.

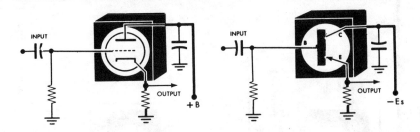

Fig. 5-7. Circuit comparison between a vacuum tube cathode follower and a transistor emitter follower.

each degree rise in operating temperature. If the temperature of the transistor increases one degree Centigrade, then 3 milliwatts is subtracted from its optimum of 150 milliwatts. Transistors have relatively low heat dissipation ratings; therefore, their power handling capabilities are limited. Transistors discussed thus far in this book have a maximum collector dissipation of 150 milliwatts. Other important specifications which are related to power ratings are: 1. Maximum permissible collector voltage; and 2. Maximum permissible collector current.

When a transistor is driven for maximum power output, it is rated as a power amplifier, although the specified maximum may be as low as 50 milliwatts; therefore, full advantage should be taken to utilize the permissible maximum allowable power—the load line must operate on the collector characteristics just within the maximum rating of the transistor—and care must be taken not to exceed these ratings. The maximum power dissipation curve for a transistor whose power handling capability is 150 milliwatts is shown in Fig. 5-10. The dissipation curve represents a safe boundary between maximum collector voltage and current.

Any point on the load line represents instantaneous values of collector voltage and current. The product of these two values represents the instantaneous power being consumed as the signal swings between maximum collector voltage and maximum collector current. The position of the load line on the collector characteristics with respect to the collector dissipation curve indicates whether or not the transistor is being operated within the safe limits specified.

For Class A circuits, the operating point represents the

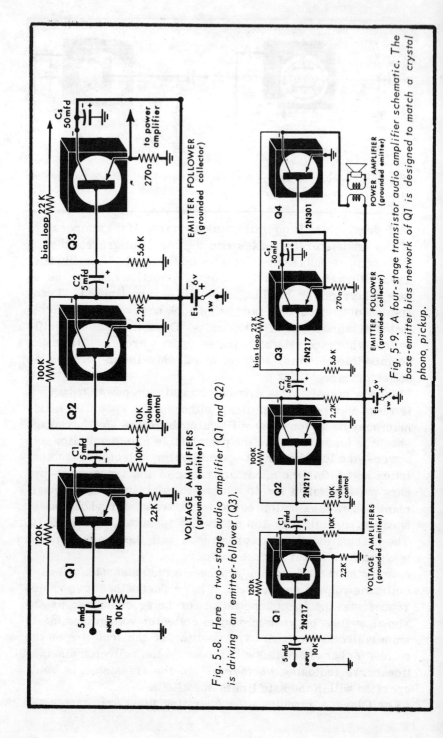

Fig. 5-8. Here a two-stage audio amplifier (Q1 and Q2) is driving an emitter-follower (Q3).

Fig. 5-9. A four-stage transistor audio amplifier schematic. The base-emitter bias network of Q1 is designed to match a crystal phono, pickup.

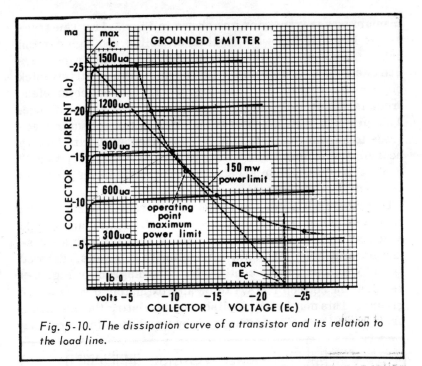

Fig. 5-10. The dissipation curve of a transistor and its relation to the load line.

maximum power being consumed while the transistor is idling (no signal). Notice in Fig. 5-10 that during the signal swing (base current) the instantaneous values of power dissipation fall well below the maximum dissipation limits. This is due to the nonlinearity of the dissipation curve; therefore, several combinations of supply voltage (Es) and load resistance (R_L) can be used without extending beyond the safe operating limits described by the dissipation curve.

The problem of heat dissipation has always confronted the transistor engineer. His prime objective has been to devise methods for better heat dispersion. The result of these endeavors has shown remarkable progress. Starting with the early units of a few milliwatts, the engineer has developed transistors capable of handling several hundred watts of power. It is evident that if the heat dissipation problem could be completely overcome, the power-handling capabilities of future transistors could be almost unlimited.

POWER TRANSISTORS

The collector characteristics for a PNP alloy-junction germanium power transistor are shown in Fig. 5-11. This transistor has a high power-handling capability and high current gain. To permit greater heat dissipation power transistors are designed with a special type housing with a relatively thick copper mounting flange. See Fig. 5-12. The collector electrode is electrically and thermally connected to the flange. This permits excellent heat transfer from the collector electrode to an external heat sink or sub-chassis. A satisfactory heat sink is achieved by bolting the transistor directly to a 6" x 6" x 1/8" aluminum sheet. Often the metal chassis serves as a satisfactory heat sink.

To further illustrate, the 2N155 power transistor has a total heat dissipation rating of 1.5 watts in free air, and a .025 watt derating per degree centigrade. When used with a satisfactory heat sink, the total dissipation is 8.5 watts, derated at .14 watt per degree centigrade. The diagram in Fig. 5-9 requires the collector of Q4 to be approximately 6 volts above ground. This may be accomplished by insulating the aluminum sheet from the main chassis, or by insulating the mounting

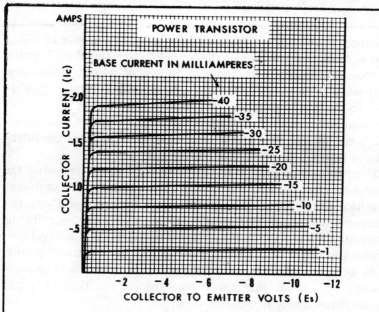

Fig. 5-11. Collector characteristics of a typical power transistor. Notice the base current bias levels are given in milliamperes and the collector current in amperes.

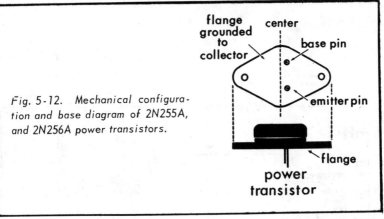

Fig. 5-12. Mechanical configuration and base diagram of 2N255A, and 2N256A power transistors.

flange grounded to collector — center — base pin — emitter pin — flange — power transistor

flange from the chassis using a .0015" mica washer or a .022" anodized aluminum washer to complete the thermal path. An important characteristic of such insulators is low thermal resistance.

SUMMARY OF BIAS CIRCUITS

Fig. 5-13 illustrates the four popular amplifier bias circuits. In circuit A the base bias is applied through R2 direct from the battery. In this arrangement the fixed base bias current remains constant regardless of a rise and fall in collector current. However, in order to provide DC stability, a resistance (R3) is connected in the emitter lead and is bypassed to prevent AC negative feedback. In circuit B notice the addition of R4 connected in series with R3. This resistance is much smaller than R3 and is used to provide a small value of AC negative feedback to improve amplifier linearity and frequency response.

In circuit C negative DC feedback is obtained by connecting the base bias resistance (R2) to the collector terminal. Notice that R2 is split to permit the use of a bypass capacitor. This prevents AC negative feedback and isolates the base and collector from direct bypass. This method is used in the secondary stage of the PA system illustrated in Fig. 6-4. In circuit D resistance R2 is halved in value and R1 is connected between the base and the positive terminal of the battery to improve feedback control. Hence, R2 provides a reverse bias. Decreasing R2 permits better feedback control; however, the increased bias current is cancelled by the reverse

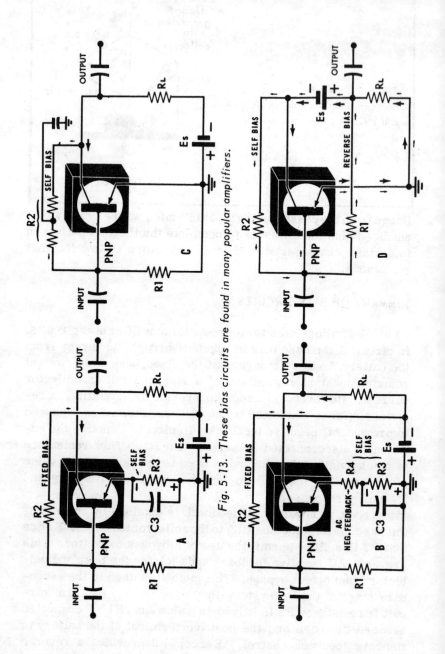

Fig. 5-13. These bias circuits are found in many popular amplifiers.

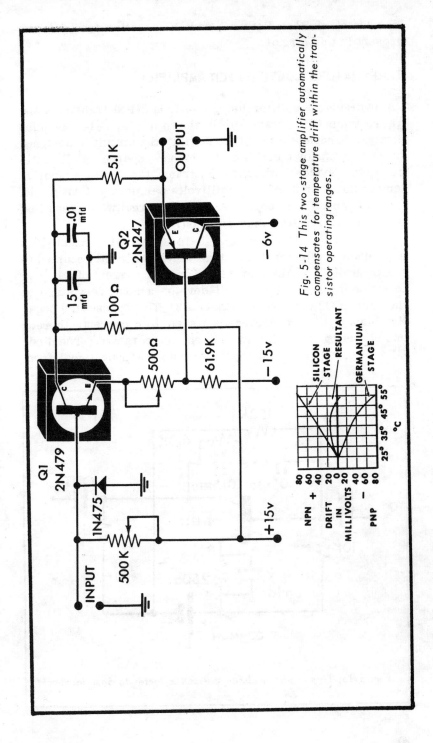

Fig. 5-14 This two-stage amplifier automatically compensates for temperature drift within the transistor operating ranges.

bias effect of R1. Both AC and DC negative feedback are present in this circuit.

TEMPERATURE-COMPENSATED AMPLIFIER

A two-stage amplifier using a silicon (NPN) transistor and a germanium (PNP) transistor is shown in Fig. 5-14. Besides having a bandpass from DC to 2 MHz, the circuit is designed to compensate for thermal drift which is inherent in all semi-conductor devices. Silicon (NPN) transistor Q1 has a positive temperature drift of 2.4 millivolts per degree Centigrade. The combined drifts oppose each other circuitwise, resulting in a minimum overall drift as illustrated.

A study of the circuit in Fig. 5-14 shows Q1 connected as a grounded-collector (AC) stage which is directly coupled to the base of Q2. Any increase in emitter current of Q1, due to a rise in temperature, will develop a positive voltage at the base of Q2 through resistance R3. This opposes the negative bias of Q2 and decreases its emitter current, thereby compensating for a rise in ambient temperature. The diode (1N475) connected to the input prevents the input signal from

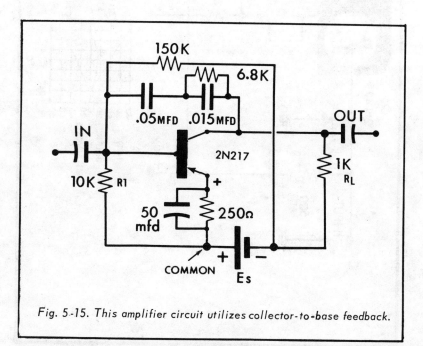

Fig. 5-15. This amplifier circuit utilizes collector-to-base feedback.

exceeding a few tenths of a volt negatively which is slightly beyond cut off for an NPN transistor. The input impedance of the amplifier is about 400K and the output is 300 ohms.

Another use of collector-to-base feedback is shown in the Fig. 5-15. This circuit provides AC negative feedback only and is used to improve the frequency response of a crystal pickup. For example, a crystal has a relatively high output impedance which is predominately capacitive; hence, when the crystal output is applied to the low input resistance of ground-emitter stage, the result is a predominance of high frequencies, aggravated by mismatch. However, with the use of a carefully designed capacitive feedback network from collector to base, the frequency response may be held flat from 50 to 15,000 Hz. The values comprising the feedback circuit are shown in Fig. 5-15. At 50 Hz, the reactance of the .05-mfd capacitor is equal to beta times R_L; therefore, as the frequency increases from 50 to 2,000 Hz the gain of the stage decreases proportionally. At 2,000 Hz the reactance of the .015-mfd capacitor is equal to the 6.8K resistor it shunts, resulting in a proportional decrease of gain at frequencies above 2,000 Hz. An emitter resistor is used to provide DC negative feedback for stability.

Chapter 6

Audio Amplifier Circuits

The efficiency of a Class A amplifier is 50%, and with a stabilizing resistor connected in the emitter lead the overall efficiency drops below 50%. Also, in single-ended amplifiers the harmonic distortion may be as high as 5%, due in part to the nonlinear rise in collector current for a linear rise in base current, which is more noticeable at higher levels of collector current.

PUSH-PULL AMPLIFIERS

For power output stages, push-pull amplifiers provide many advantages, such as cancellation of even harmonics and the effects of DC saturation on transformer linearity. A typical Class A push-pull audio amplifier using two PNP transistors is shown in Fig. 6-1. The DC base biasing for each transistor is determined in the same manner as Class A single-ended stages. However, in a Class A amplifier the average collector current flows constantly with or without an applied signal, and this constitutes a continuous drain on the power supply which is usually a battery.

Greater efficiency can be obtained by using two Class B amplifiers in push-pull for the power output stage. In such a circuit each transistor is biased to near cut-off, resulting in a low collector current which is ideal for conserving power during the idling periods. Class B push-pull transistor amplifiers are simple to construct, consume very little standby power, have an operating efficiency up to 78%, and are used to a greater extent than Class B vacuum tube amplifiers. In Class B operation, transistors can be connected as grounded-emitter, grounded-base, or grounded-collector stages, and are usually coupled to a driver by a transformer or direct connection.

A transformer-coupled Class B push-pull amplifier circuit is shown in Fig. 6-2. A low value of forward bias (near cut-off) is applied to the base of each transistor through a voltage-divider network consisting of R12, R13, R14, and R15. See Fig. 6-3. To compensate for any change in the low DC operating point due to temperature, R15 may be a thermistor or some other temperature-compensating device.

MOBILE PUBLIC ADDRESS SYSTEM

A complete transistor PA system is shown in Fig. 6-4. This high-power amplifier is designed for compactness and dependability and operates directly from a 12-volt storage battery. The system incorporates some interesting circuit features, including two 2N301 power transistors operated in Class B push-pull, capable of handling 10 watts of audio power. The complete system is explained in the following stage-by-stage description.

Q1—Preamplifier

This stage operates as a preamplifier and is driven by a dynamic microphone through coupling capacitor C1. Base bias is obtained through a voltage divider consisting of resistors R1 and R2.

The collector circuit is connected to the negative terminal of the supply battery Es through load resistor R3 and de-coupling resistor R16. In addition to the decoupling function, R16 also serves as a dropping resistor to reduce the collector voltage, which also aids in the reduction of noise.

Decoupling resistor R16 operates in conjunction with bypass capacitor C6 located in the collector circuit and functions as a filter to prevent feedback from the common supply line, which would result in motorboating. Notice the output signal is capacitively-coupled through C2 and volume control R5 to the following stage (Q2).

Q2—Intermediate Amplifier

This stage functions as the second audio or intermediate

amplifier. The bias loop from the collector to the base uses two 100K resistors, R4 and R6, with bypass capacitor C4 connected at the center. The purpose of C4 is to bypass the AC component in the bias loop to ground. The split resistor network R4 and R6 used in conjunction with C4 isolates the input and output circuits. Therefore, only DC negative feedback is fed to the base from the collector. The output signal

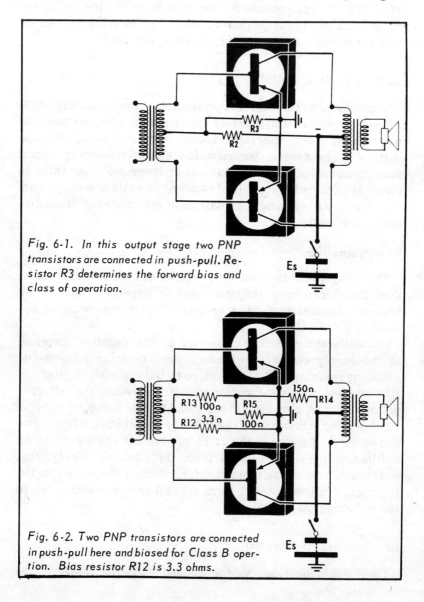

Fig. 6-1. In this output stage two PNP transistors are connected in push-pull. Resistor R3 determines the forward bias and class of operation.

Fig. 6-2. Two PNP transistors are connected in push-pull here and biased for Class B operation. Bias resistor R12 is 3.3 ohms.

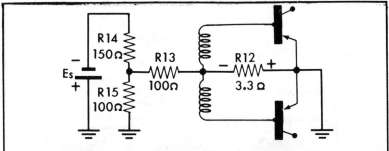

Fig. 6-3. Forward bias network for Class B push-pull. The voltage drop across R12 provides the forward bias for both transistors. A thermistor may be substituted for R15. This would be necessary if the amplifier is subjected to extremes of temperature.

of this stage is capacitively-coupled through C3 to the following stage (Q3).

Q3—Emitter Follower

This stage operates as an emitter-follower which permits direct coupling to the driver stage. The output signal is then taken from the emitter at R10, which is connected directly to the base of the driver stage (Q4). Since an emitter-follower does not invert the phase, its base bias and negative feedback is supplied from the Q4 collector through the R9 bias loop.

Q4—Audio Power Driver

This stage uses a 2N301 power transistor operating as a grounded-emitter amplifier. Base bias is obtained through emitter resistor R10 of Q3. Notice that the Q4 collector also supplies base bias and negative feedback to Q3; and since this feedback is over two stages it becomes unnecessary to stabilize Q4 separately. The output of this stage is transformer-coupled to the push-pull power amplifier consisting of Q5 and Q6, and decoupled from the common supply line by C5 and R11.

Q5 and Q6—Push-Pull Power Output

The push-pull output stage uses two power transistors, type

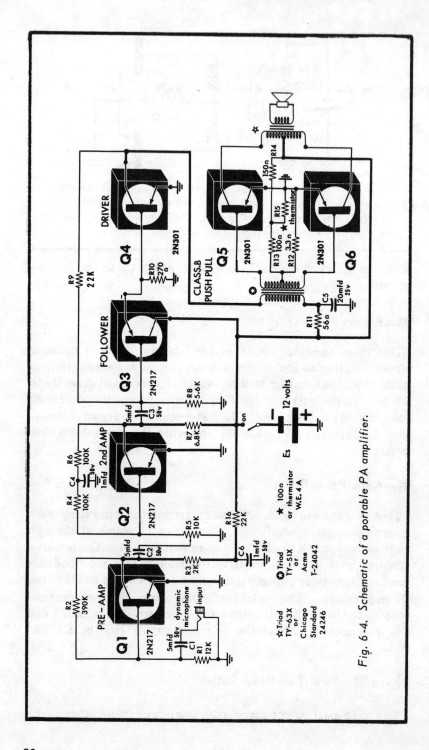

Fig. 6-4. Schematic of a portable PA amplifier.

2N 301, operated as grounded-emitter amplifiers and biased for Class B operation. Base bias for both transistors is supplied by resistor R12, which is 3.3 ohms. The voltage drop across this resistor provides a low value of forward bias and is sufficient to avoid cut-off, thus preventing distortion during crossover between the positive and negative swings of the input signal. See Fig. 6-5.

In biasing transistors for Class B operation the collector current should not be cut off, due to the nonlinearity of collector characteristics at low current levels. However, the base bias must be the correct value to provide a balanced output. Using a sine-wave input and viewing the output with an oscilloscope, the bias should be adjusted for minimum distortion. See Fig. 6-5. When the forward bias between the base and emitter is zero, the collector current is practically cut off, with the exception of Ico. This differs from the vacuum tube where the grid must be driven sufficiently negative to cut off plate current.

Resistors R12, R13, R14, and R15 (Fig. 6-4) form a voltage divider network across the battery for good regulation. To minimize the effect of temperature changes, a thermistor may be used in place of R15. This substitution is necessary only if the amplifier is subjected to extremes of temperature.

Fig. 6-6 shows another type of push-pull amplifier, driven by a capacitively-coupled phase inverter, thus eliminating the need for an input transformer. Two crystal diodes are used to shunt the base input circuits; these diodes prevent each transistor from being biased beyond cut-off by the positive peaks of the input signal.

Each transistor is driven alternately into conduction by the negative signal swings of the input signal (forward bias PNP). During positive signal swings each diode conducts and short circuits these peaks, thereby preventing a large reverse bias from being applied to the base of each transistor. Without diodes the positive peaks (reverse bias PNP) would hold the transistor beyond cut-off too long, resulting in distortion. See Fig. 6-7.

COMPLEMENTARY PUSH-PULL

Although the circuits discussed thus far have shown a leaning to PNP type transistors, it is possible to combine PNP

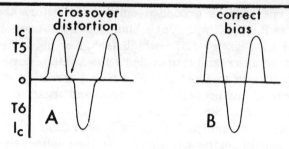

Fig. 6-5. Curve A illustrates distortion due to incorrect base bias. B illustrates balanced output obtained when forward bias is correctly adjusted.

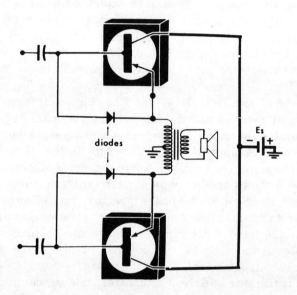

Fig. 6-6. Push-pull output stage using capacitor input coupling.

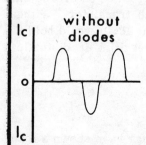

Fig. 6-7. Illustrated here is the distortion of the output signal when shunt diodes are omitted in Fig. 6-6.

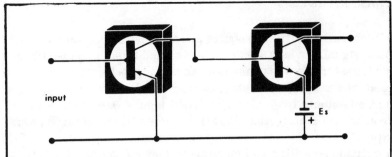

Fig. 6-8. *The complementary symmetry of PNP and NPN transistors is illustrated in this single-ended amplifier circuit. Notice direct coupling.*

and NPN transistors in the same circuit. The complementary symmetry of PNP and NPN types permits their combined operation in circuits not possible with vacuum tubes. For instance, the cathode of a vacuum tube emits electrons similar to the emitter of an NPN transistor. The novel feature of a PNP transistor is that it emits positive charges (holes) —a feat not possible in a vacuum tube. Combining the opposite features of these two types of transistors, it is possible to design circuits that have no vacuum tube equivalents.

For example, a Class B push-pull amplifier can be constructed without the use of transformers or a phase inverter, thanks to the complementary symmetry of PNP and NPN transistors. The PNP transistor will amplify the negative portion of the input signal, while the NPN will amplify the positive portion. The two portions of the signal are then combined in a common output, which may be a voice coil. Fig. 6-8 shows a practical arrangement where an NPN transistor is directly coupled to a PNP transistor. The overall gain of this type of amplifier is relatively low, compared to transformer coupling, due to mismatch, but the simplicity of the circuit and minimum number of components overcome the gain disadvantage. Although the amplifier in Fig. 6-8 is single-ended, it also can be used in push-pull as shown in Fig. 6-9. In low-power amplifiers this complementary push-pull circuit is capable of producing 400 milliwatts output using four transistors each with a 100-milliwatt rating. The driver transistors in Fig. 6-9 could have a power rating less than those used for the output stage.

PHASE INVERTERS

Push–pull amplifiers require two equal but oppositely–phased input signals, which are usually obtained from a push–pull input transformer or a phase inverter. Fig. 6-10 shows a split-load phase – inverter stage which supplies two oppositely – phased outputs from a single-ended input. One input is taken from the collector, the other from the emitter. Actually, this type of phase inverter combines the phase characteristics of a grounded-emitter and emitter-follower stages.

When collector and emitter resistors R3 and R4 are equal, the stage provides a nearly balanced push-pull output signal. A slight unbalance is to be expected, since the emitter current through R4 is slightly larger than the collector current flowing through R3. A balance may be accomplished by making the resistance of R3 slightly higher than R4. However, when this type of phase inverter is used to drive a Class B push-pull stage some unbalance will exist, due to the alternate loading of R3 and R4 by each half cycle of the input signal. For example, a change of load on the collector resistor will affect only the collector output, while a change of load on the emitter resistor affects both outputs, due to emitter feedback. To overcome this problem R3 and R4 are made small, and a resistance added in series with each connection to the push-pull stage. This tends to make the driver impedance constant and provide the necessary decoupling.

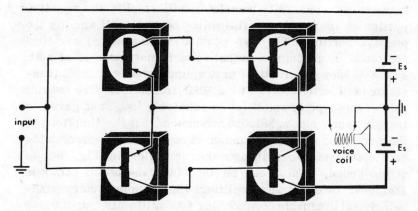

Fig. 6-9. Circuit showing two PNP and two NPN transistors directly coupled in push -pull. Notice the absence of input and output transformers.

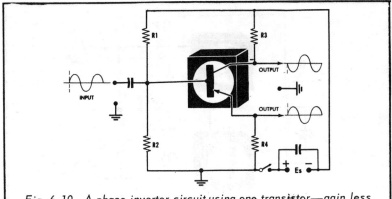

Fig. 6 -10. A phase inverter circuit using one transistor—gain less than unity. This stage makes possible the elimination of an input transformer.

Fig. 6-11 shows a phase inverter using two NPN transistors connected as grounded-emitter stages. The signal input to the base of Q1 causes emitter current to flow through R1. The voltage developed across R1 drives Q2 in the opposite phase. Fig. 6-12 is another type of phase inverter circuit. The signal applied to Q1 is amplified and appears inverted at the collector of Q1. A sample of this signal is coupled to the base of Q2 through R3 and C3. Q2 amplifies the sample signal and it appears at the collector of Q2 equal in amplitude but opposite in phase.

COMPLEMENTARY-SYMMETRY PUSH-PULL AMPLIFIERS

An NPN and a PNP transistor connected in complementary-symmetry, operating as a common-emitter push-pull stage, is shown in Fig. 6-13. When the two base electrodes are driven positive by a sine-wave signal, transistor Q1 (NPN) conducts and delivers a half sine wave of current through R_L. When the two base electrodes are driven negative, transistor Q2 (PNP) conducts and delivers a half sine wave of current through R_L in the opposite direction to that of Q1. No appreciable DC current can pass through R_L if the two transistors are balanced. However, a DC current passes through the two transistors which are in series across the total power supply, and the drain on this supply increases only during signal drive. The advantage of this type of circuit is that when Q1 con-

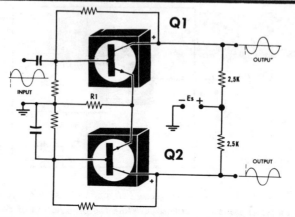

Fig. 6-11. A phase inverter circuit using two NPN transistors. Gain greater than unity.

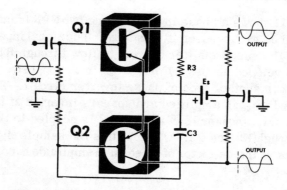

Fig. 6-12. In this phase inverter circuit a portion of the output signal of Q1 is coupled to the base of Q2 through R3, C2. This **small** signal is then amplified and inverted.

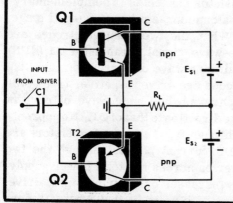

Fig. 6-13. This push-pull circuit uses a PNP and an NPN in a common-emitter output stage.

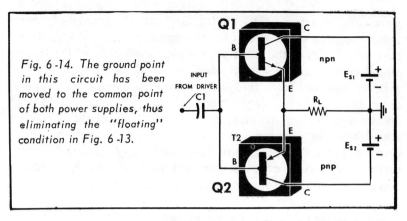

Fig. 6-14. The ground point in this circuit has been moved to the common point of both power supplies, thus eliminating the "floating" condition in Fig. 6-13.

ducts, the electron current passes out of the base electrode, and at Q2 the electron current passes into the base electrode; therefore, no DC base current is required to establish the critical operating point required for Class B push-pull stage. The disadvantage of this circuit is that the power supply is floating at the signal end of output load resistor R_L. If the ground point of this stage is changed from the emitters to the common point of both power supplies, the circuit is converted to a common-collector push-pull stage. By doing this the disadvantages mentioned are eliminated. See Fig. 6-14.

Although the power gain of this circuit is much less that of the grounded-emitter type, it offers the following advantages:

1. The advantages of negative feedback
2. Reduced distortion
3. Reduced phase shift at high frequencies
4. Establishes a better balance
5. Low impedance output for voice coil

Fig. 6-15 shows a complementary-symmetry, push-pull circuit similar to that in Fig. 6-14, but revised to operate on a single power supply. This is accomplished by inserting a blocking capacitor (C2) between the emitter and resistor (F_L).

SUMMARY OF MAJOR STAGES

With an oscilloscope adjusted to show three positive peaks and two negative peaks to clearly illustrate phase inversion, the common-emitter circuit in Fig. 6-16 reveals the accompanying waveforms. Notice that the collector waveform peaks are inverted, thus conforming with the theory of wave inversion. In the common base-circuit in Fig. 6-17A notice that the polarity of the positive and negative peaks of the input signal remain unchanged, indicating no change in phase. In making these two tests the gain of the oscilloscope was attenuated 10 to 1 when viewing the collector waveforms. In the common-collector circuit (Fig. 6-17B) the gain of the oscilloscope was not changed. Notice that the input signal was not inverted but a voltage gain of less than unity resulted. The three tests prove conclusively that each circuit has characteristics similar to its vacuum tube counterpart.

PRECAUTIONS ON USING TRANSISTORS

<u>Mechanics</u> and <u>Installation</u>: Since transistors are mechani-

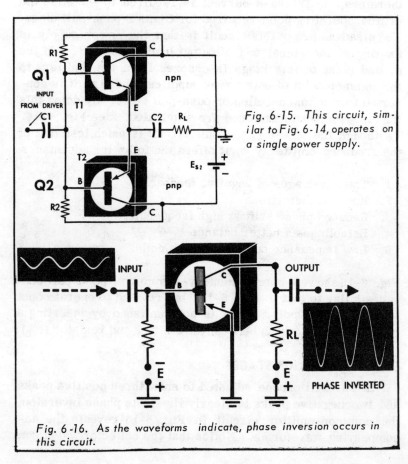

Fig. 6-15. This circuit, similar to Fig. 6-14, operates on a single power supply.

Fig. 6-16. As the waveforms indicate, phase inversion occurs in this circuit.

cally rugged devices, undue concern about reasonably rough treatment is not necessary. The primary consideration is the avoidance of extreme shock and excessive bending or twisting of the leads. Power transistors should be mounted on a suitable radiator or heat sink. In circuits where the collector is above ground, the radiator should be insulated from the chassis with a mica washer and bolted to the chassis with insulated bolts.

Temperature: Temperature extremes can have a severe effect on transistor life. Permissible storage temperatures are well within the range normally encountered, but over-heating in a circuit can be disastrous, especially in power amplifiers, as mentioned above. Heat increases collector cut-off current (Ico) which, in turn, reduces power output and further increases the heat developed. This may result in a "runaway" condition. The circuit can be stabilized by using a thermistor in the base circuit; thus, a rising temperature decreases the base-to-emitter voltage and stabilizes Ico. Precautions should be observed while soldering. It is best to solder with the transistor out of the socket; or, if the transistor has flexible leads, hold pliers on the lead between the point being soldered and the transistor.

Electrical Conditions: Transistors are capable of extremely long life if operated within their established ratings. However, small excesses in the voltage or power ratings may destroy the transistor instantaneously. Therefore, precaution is necessary when experimenting or testing new circuits. Some important considerations follow:

First: Double check the polarity of the supply voltage. Incorrect polarity endangers both transistors and electrolytic capacitors.

Second: When first testing a new circuit, apply voltage in easy stages, starting at a lower than normal value to see if operation appears to be normal. Beware of high-voltage surges. It is a good idea to load a 6- or 12-volt unregulated electronic power supply with a storage battery to stabilize the voltage.

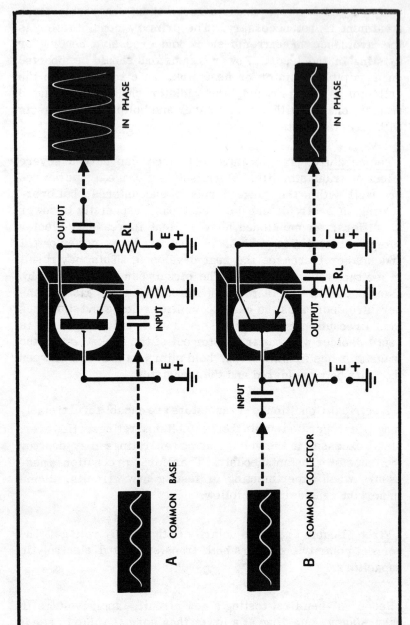

Fig. 6-17. There is no phase inversion in the common-base circuit (A) or in the common-collector circuit (B).

Third: Before the circuit is allowed to operate for an extended period, measure the collector current in the power stages and adjust bias if necessary.

Fourth: Never operate a power stage without connecting a load to the output.

Chapter 7

RF and IF Amplifiers

Several methods of coupling RF and IF stages are shown in Fig. 7-1. Diagram 1 shows a transformer coupling with a tuned primary and untuned secondary. The tuned primary circuit represents a high impedance capable of matching the collector output impedance. The low-impedance untuned secondary is suitable for matching the base input impedance of the following stage. Diagram 2 shows a resonant circuit formed by two fixed capacitors, C1 and C2, and a permeability-tuned inductance. The ratio of these two capacitances determines the impedance stepdown. For example, the greater the capacity of C2 in relation to C1 the lower its impedance.

In diagram 3 in Fig. 7-1 coupling is achieved with stepdown impedance taps on a single tuned circuit. Notice the low-impedance tap is capacitively coupled to the base input of the following stage. In Diagram 4 impedance taps for the collector and input circuits permit the unloading of the tuned circuit and thereby maintain high selectivity. Diagram 5 shows a doubled-tuned IF transformer. The tuned primary is tapped to match the lower impedance of the base input impedance. This arrangement permits adjustment of the effective coupling to provide the desired response curve.

TRANSISTOR RADIO RECEIVERS

A schematic diagram of a simple radio receiver is shown in Fig. 7-2. The unit consists of a tuned circuit and a crystal detector, capacitively coupled to a transistor audio amplifier. Fig. 7-3 is the schematic diagram of a typical tran-

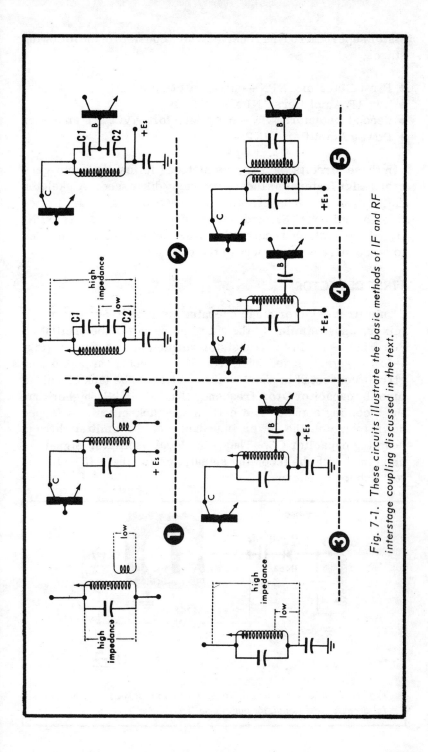

Fig. 7-1. These circuits illustrate the basic methods of IF and RF interstage coupling discussed in the text.

sistor superheterodyne receiver which consists of the following stages:

- First Detector: NPN—mixer-oscillator
- Two IF Amplifiers: NPN —455 kHz
- Second Detector: NPN—peak detector, AVC, AF amplifier
- Power Amplifier: PNP — Class A

In these circuits the DC bias methods follow the same general design features as those previously discussed. A skeleton diagram of the bias circuits is shown in Fig. 7-4. Notice that the mixer, IF, and second detector stages use NPN transistors which require the base and collector electrodes to be positive with respect to the emitter.

FIRST DETECTOR

This first detector stage operates as a self-oscillator and mixer, and is similar to the autodyne vacuum tube oscillator. See Fig. 7-5. In the transistor version of this circuit (Fig. 7-5B) a signal is fed back from the collector through a tank circuit connected in shunt to the emitter. The tank circuit is tuned by capacitor C to a frequency that is 455 kHz higher than the incoming signal. The coil in the tank circuit is tapped down to match the lower impedance of the emitter through coupling capacitor C3. Thus, the local oscillator signal is combined with the incoming signal, producing a 455-kHz intermediate frequency.

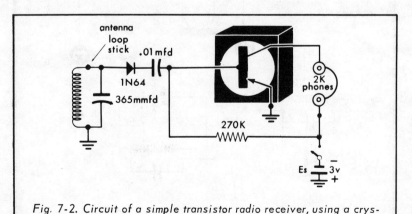

Fig. 7-2. Circuit of a simple transistor radio receiver, using a crystal detector and transistor audio amplifier

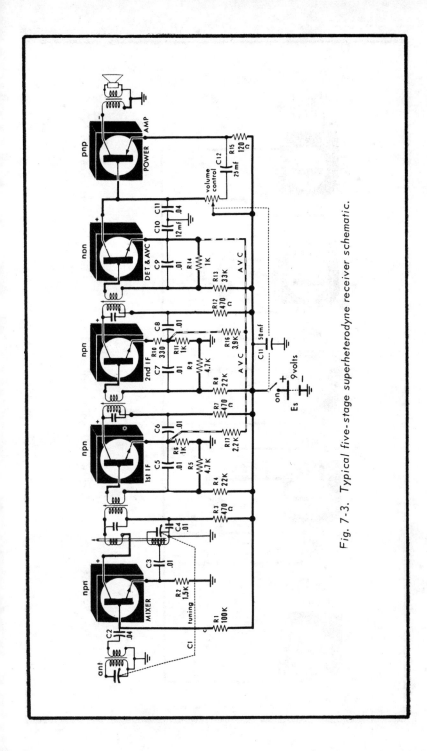

Fig. 7-3. Typical five-stage superheterodyne receiver schematic.

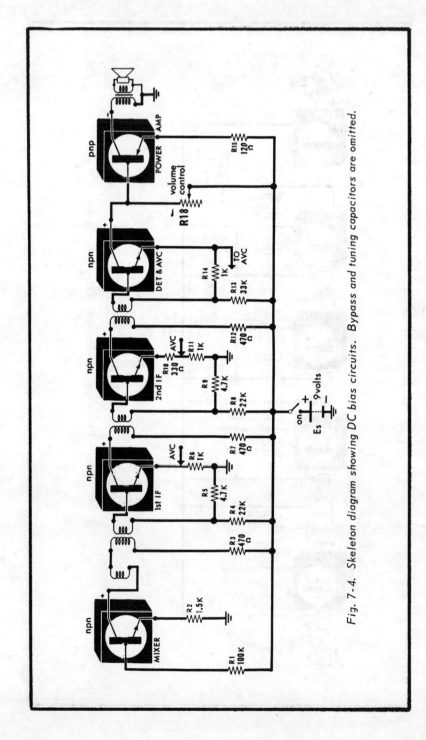

Fig. 7-4. Skeleton diagram showing DC bias circuits. Bypass and tuning capacitors are omitted.

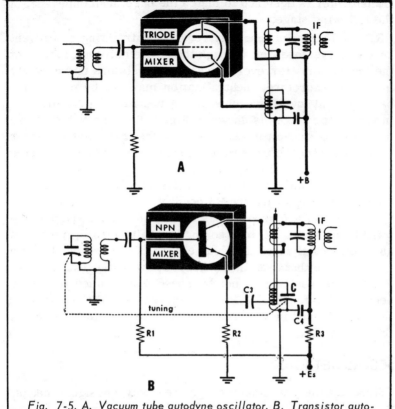

Fig. 7-5. A. Vacuum tube autodyne oscillator. B. Transistor auto-dyne oscillator.

The fixed base bias of this stage is obtained from the positive terminal of Es, through the base bias loop consisting of R1 and R2. Resistor R2, located in the emitter lead, functions as a stabilizer. The bias voltage developed across the base-emitter junction of this stage is approximately .03 volt. Resistor R3, located in the collector circuit, functions as a decoupling resistor to prevent undesired feedback from the common supply line.

IF STAGES

The two IF stages are tuned to 455 kHz. The primary of each IF transformer is tuned with a fixed capacitor and permeability-tuned. Each primary is tapped to match the output impedance of the collector, whereas the secondary of each transformer is designed to match the base input impedance of the following stage.

RF and IF stages generally require neutralizing to prevent regenerative feedback through the interelectrode capacity of the transistor. However, in some transistors this capacity is small; therefore, neutralization may not be necessary, especially at low IF frequencies. A neutralizing circuit used in transistor stages is shown in Fig. 7-6. The value of the neutralizing capacitor depends upon the transistor collector capacitance and the turns ratio of the IF transformer connected to the output side of the transistor.

The fixed bias for the first IF stage utilizes a voltage divider consisting of R4 and R5 connected across Es. See Fig. 7-4. The voltage drop across R5 provides the required forward bias, while R6 located in the emitter lead, functions as a DC stabilizer. The bias circuit of the second IF stage is similar to that used in the first IF, with the exception of R10. This resistor is not bypassed and is used to supply negative AC feedback. Both IF stages are AVC-controlled; the system used in this receiver will be discussed later in this Chapter.

SECOND DETECTOR

Since the second detector operates only on signal peaks, it is biased close to cut-off, resulting in a relatively low value of collector current (see Fig. 7-3). This provides rectification and amplification of the input signal. In Fig. 7-4 it will be seen that the forward bias of this stage is pro-

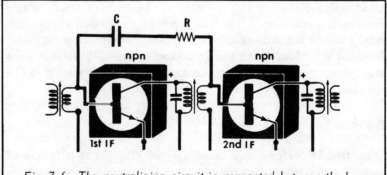

Fig. 7-6. The neutralizing circuit is connected between the base electrodes of IF stages.

vided by the voltage drop across R14, which is a 1K resistor. This resistor forms a voltage divider with R13, which is 33K, and this provides a forward bias of approximately .15 volt. An NPN transistor requires a positive voltage at its collector. However, the drop across R18 is such that the upper end becomes negative with respect to the arm, and this provides the proper negative polarity for direct coupling to the base of the PNP power amplifier.

The amplitude of the audio signal applied to the output stage depends upon the setting of the volume control. For example, as the arm of the control is moved towards the bottom of R18, the volume increases. At the same time the by-pass action of C12 (Fig. 7-3) across emitter resistor R15 is made more effective. Notice that C12 is connected in series with the volume control and functions as a tone compensator by introducing selective degeneration. When the arm of the control is moved towards the top of R18 the volume decreases due to the shunting effect of C12. Also, the bypass action of C12 across R15 is made less effective due to the increase of the series resistance of the volume control. This maintains good tonal quality at all settings of the volume control.

AUTOMATIC VOLUME CONTROL

The purpose of automatic volume control is to maintain a constant level of volume. This is generally accomplished by applying DC feedback from the second detector to one or more RF and IF stages.

The average value of a carrier wave is zero, regardless of its amplitude. However, when it is rectified and filtered it produces a DC voltage which is proportional to the amplitude of the carrier. See Fig. 7-7. In vacuum tube receivers this DC voltage is usually negative and is generally obtained from the output of the second detector. It is then applied through a filter circuit to the control grid of one or more RF and IF stages.

Transistors require current feedback for RF and IF gain control which is usually obtained from the output of the second detector. This feedback current is passed through a filter circuit to remove the audio component and is then applied to the emitter or base electrodes of one or more RF and IF stages. Fig. 7-8 shows the AVC system used in the receiver just discussed. Notice the emitter current of the detector

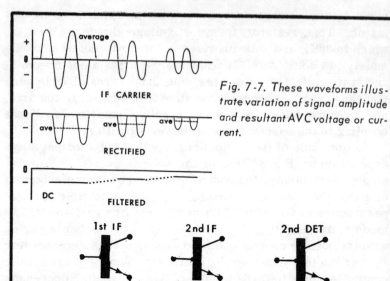

Fig. 7-7. These waveforms illustrate variation of signal amplitude and resultant AVC voltage or current.

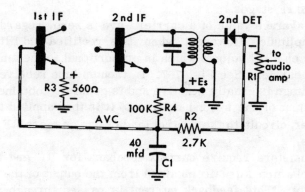

Fig. 7-8. AVC emitter bias control circuit. The AVC supply bias is represented by dashed lines.

Fig. 7-9. AVC base bias control circuit. Notice that initial bias is applied to the detector diode through R4.

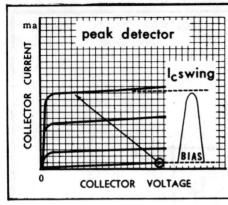

Fig. 7-10. The arrow indicates the base current swing and the resultant swing of collector current of a peak detector.

flows through emitter resistors R6 and R11 of the first and second IF stages. The operation of this circuit maybe described as follows:

1. An increase in signal carrier strength will cause an increase of the emitter current of the peak detector.
2. This current flows in a parallel path formed by emitter resistor R6 and R11 of the first and second IF stages. Series resistors R16 and R17 are used to limit the amount of AVC current applied to each stage.
3. An increase in current through R6 and R11 will develop a positive voltage at the emitter of each IF stage.
4. When a positive voltage is applied to the emitter of an NPN transistor, it will tend to oppose the forward bias and lower the gain, thereby compensating for an increase in signal strength.
5. A decrease in signal strength will lower the AVC current, which will aid the forward bias and cause an increase in gain of both IF stages.

The system requires appreciable AVC power to provide a wide control range, and this is accomplished by using a peak detector. In receivers using crystal diode detectors, some means of AVC current amplification is required. This is achieved by connecting the AVC feed line to the base electrodes of the stages to be controlled, where a small change in AVC current will produce a relatively large change in the emitter current of each stage under control. In this manner the controlled stages function as their own AVC amplifiers. A circuit illustrating this type of control is shown in Fig. 7-9; its operation is described as follows:

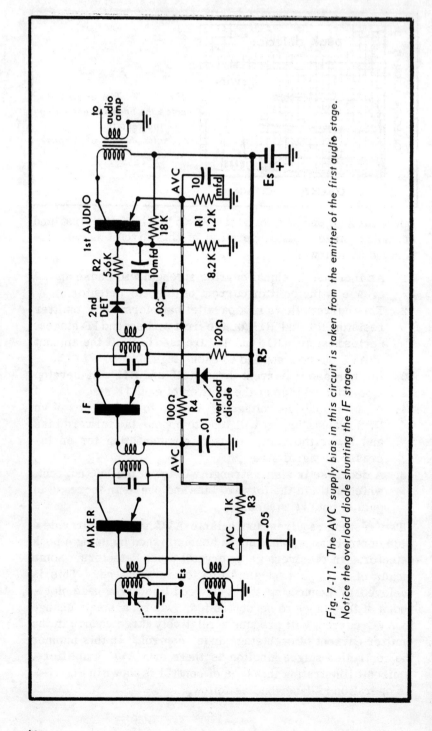

Fig. 7-11. The AVC supply bias in this circuit is taken from the emitter of the first audio stage. Notice the overload diode shunting the IF stage.

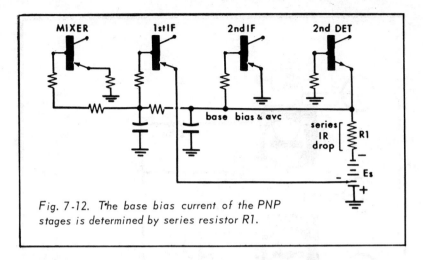

Fig. 7-12. The base bias current of the PNP stages is determined by series resistor R1.

1. The crystal diode detector is connected to provide a negative potential above ground which is applied to R1 and then to the base of the first IF stage.
2. An increase in signal carrier strength will cause an increase in the negative potential at the top of R1 and the base electrode of the first IF.
3. A negative potential applied to the base of an NPN transistor will oppose the forward bias, which in turn decreases the emitter current and also the gain of the stage, whereas a decrease in signal strength will aid the forward bias and increase gain.

Referring to the diagram in Fig. 7-9 it will be seen that the diode detector is biased slightly in the forward direction by the positive supply voltage through R2 and R4. This shifts the operating point of the diode detector to a more efficient portion of its characteristic curve (see Fig. 7-10).

Still another method of AVC is shown in Fig. 7-11. Here the AVC voltage is taken off the emitter resistor (R1) of the first audio stage. This is possible since the diode detector is DC coupled to this stage by resistor R2. (If capacitively coupled the DC component of the rectified signal would be lost.) This amplifiers the AVC voltage which is then applied through the filter network to the base electrode of the mixer and IF stages. Here is how this system operates:

1. The diode detector is connected to produce a positive

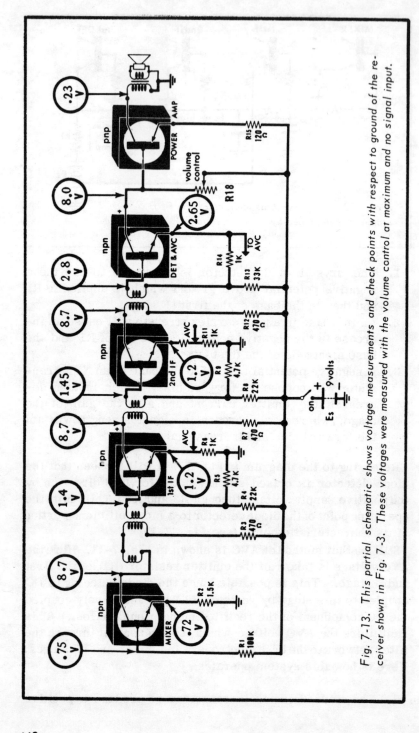

Fig. 7-13. This partial schematic shows voltage measurements and check points with respect to ground of the receiver shown in Fig. 7-3. These voltages were measured with the volume control at maximum and no signal input.

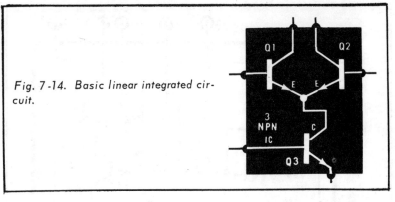

Fig. 7-14. Basic linear integrated circuit.

voltage at the base of the first audio which is the average DC value of the rectified carrier. See Fig. 7-7.

2. An increase in signal strength produces an increase of positive voltage at the base of the first audio stage.

3. When a positive voltage is applied to the base of a PNP transistor, it reduces the emitter current, thereby reducing the negative voltage at the emitter.

4. Since the two controlled stages are PNP transistors, a reduction of negative voltage at the base electrode decreases the forward bias, which in turn decreases the gain. This compensates for the increase of signal at the detector, whereas a decrease in signal strength will produce an opposite effect.

Referring to Fig. 7-11 it will be seen that a diode is connected in parallel with the output of the IF stage. On a normal signal the collector current develops a small voltage across R5. The polarity of this voltage is such that it applies a reverse bias across the diode. With this condition the diode represents a high impedance and does not present a load to the tuned circuit. However, when the signal increases in amplitude, the voltage swing across the tank circuit develops a forward bias and causes the diode to conduct. This in turn loads the tuned circuit and attenuates the signal.

Fig. 7-12 shows another method of providing gain control. In this system the emitter of the second detector, an NPN transistor, is biased negatively by the battery through R1. This resistor is in series with the base bias supply of the mixer and IF stages, and operates as an automatic volume control. This function is described as follows:

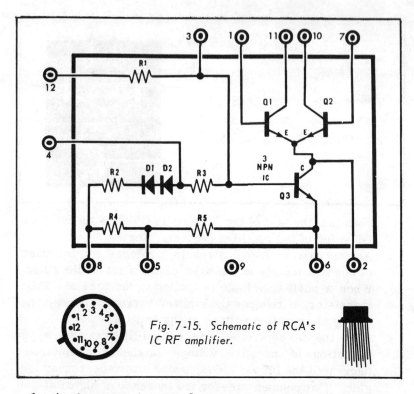

Fig. 7-15. Schematic of RCA's IC RF amplifier.

1. An increase in signal carrier strength will develop an increase in the emitter current of the second detector, which in turn increases the voltage drop across R1.
2. The increased voltage developed across R1 opposes the forward bias applied to the mixer and IF stages. This in turn decreases the gain of these stages. For a decrease in signal carrier strength the gain of these stages is increased.
3. The emitter of the first IF stage being connected to a lower fixed negative bias gives this stage more effective gain control.

The base, emitter, and collector voltages given in Fig. 7-13 were measured with a VTVM at the points indicated. (Volume control at maximum and no input signal.) Resistance measurements should not be made while the transistors are connected in the circuit.

Transistors provide a long and trouble-free life. However, should a transistor be suspected, a quick check may be made by substituting a good unit. A quick check to ascertain the condition of a transistor if removed from circuit is given in Fig. 2-4. This shows a simple method for checking both sections. In Fig. 2-4A the base electrode is connected to the negative test lead of the ohmmeter, while the positive lead is switched from the emitter to the collector. Both readings will indicate forward resistance (PNP). Fig. 2-4B shows the base electrode connected to the positive lead of the ohmmeter, while the negative test lead is switched from the emitter to the collector. Both readings will indicate reverse resistance (PNP). For NPN transistors the forward and reverse resistance indications are the opposite to those shown in Fig. 2-4.

RF AND IF INTEGRATED CIRCUITS

A significant advance in the field of semiconductor technology was the introduction of "linear integrated RF and IF amplifier circuits" (LIC). This mode of operation proved very successful and was made popular by the selection of the basic differential amplifier with an additional NPN stage. See Fig. 7-14. These three integrated active components, Q1, Q2, and Q3, may be connected to perform any one of the following functions and fabricated on a monolithic block to contain resistors and diodes. See Fig. 7-15:

- RF amplifier
- Oscillator
- Mixer
- IF amplifier with AGC
- Autodyne converter
- Video amplifier TV

The versatility of this multi-purpose integrated package is further enhanced since it allows the choice of the following amplifier configurations:

- Common-emitter
- Common-base
- Common-collector

This simplifies input and output matching between stages. A basic differential amplifier (LIC) is illustrated in Fig. 7-16.

Transistor Q1 and Q2 form the actual differential pair, and Q3 operates as a constant-current supply. Adjusting the base bias voltage of Q3 establishes the proper collector current for Q1 and Q2. This constant-current supply feature is referred to as a "constant-current sink." The base bias voltages (V1 and V2) applied to Q1 and Q2 determine the value of collector current in both stages. The two currents I_{c1} and I_{c2}, should be equal for linear operation; therefore, Q1 and Q2 should be a near matched pair.

LINEAR INTEGRATED RF AMPLIFIER

In Fig. 7-16 the input signal is applied to the base of Q1, and its collector is connected directly to the supply voltage Vcc. The emitter of Q1, which is the output, is connected to the emitter of Q2. (Input to base; output from emitter.) This circuit configuration makes Q1 a "common-collector" amplifier, with its high current gain, high input resistance, and relatively low output resistance. The base of Q2 is at AC ground potential due to the very low reactance of the bypass capacitor. The emitter input of Q2 is connected directly to the emitter output of Q1. The collector output of Q2 is connected to the tuned load. This circuit configuration makes

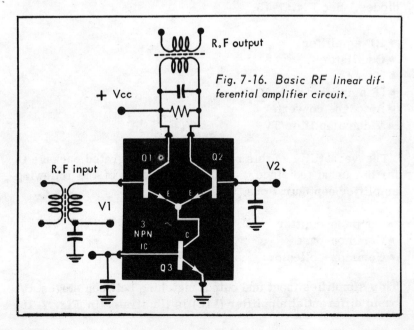

Fig. 7-16. Basic RF linear differential amplifier circuit.

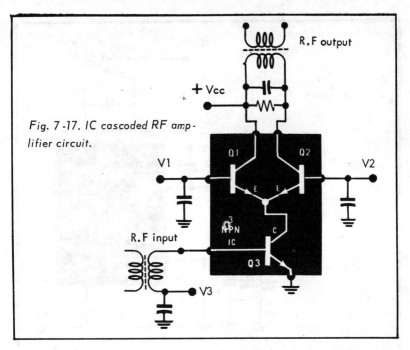

Fig. 7-17. IC cascoded RF amplifier circuit.

Q2 a common-base stage, with its high voltage gain, low
input resistance, and relatively high output resistance. In
summing up, Q1 and Q2 constitute a two-stage RF amplifier
where a common-collector stage drives a common-base stage.
Q3, as stated previously, serves as a constant-current
supply; i.e., any variation in voltage between the collector
and emitter of Q3 will not effect its collector current; hence,
a constant-current supply to both Q1 and Q2.

Three other methods of operation using this integrated package are shown in Figs. 7-17, 7-18, and 7-19. Fig. 7-18
shows the three transistors connected for mixer operation.
In this configuration all three stages (Q1, Q2, and Q3) operate
as common-emitter amplifiers. The received signal is applied
in push-pull to Q1 and Q2, while the local oscillator signal is
applied to the base of Q3. Hence, the collector current of Q1
and Q2 contains the oscillator signal. This system has many
advantages over other types of mixers. For example, the
differential amplifier operates similar to a balanced modulator;
i.e., in the absence of an RF signal the oscillator signal is
cancelled, thus preventing the oscillator signal from entering
the IF amplifier during the absence of an RF signal. Another
feature is the cancellation of even harmonics. The resistor

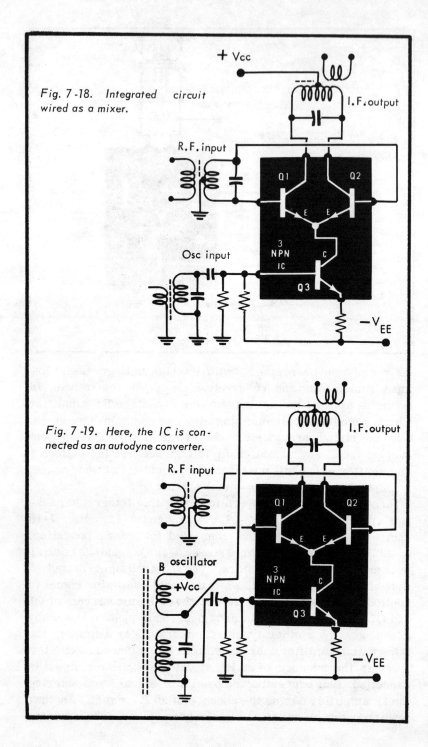

Fig. 7-18. Integrated circuit wired as a mixer.

+Vcc

I.F. output

R.F. input

Q1   Q2

E   E

3 NPN IC

Osc input

C

Q3

−V_EE

Fig. 7-19. Here, the IC is connected as an autodyne converter.

I.F. output

R.F input

Q1   Q2

E   E

3 NPN IC

B oscillator
+Vcc

C

Q3

−V_EE

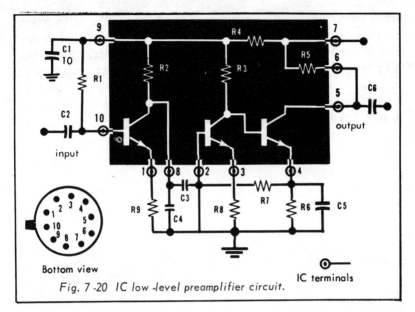

Fig. 7-20 IC low-level preamplifier circuit.

in the emitter lead of Q3 improves circuit linearity and provides a fairly good oscillator sinewave. The more sinusoidal the oscillator signal entering the mixer stage the better the rejection of spurious responses.

CASCODED RF AMPLIFIER (LIC)

Fig. 7-17 shows the integrated package connected as a cascoded RF amplifier. In this configuration the RF input is applied to the base of Q3, operating as a common-emitter stage in cascode with Q2, which is connected to operate as a common-base amplifier. Notice the base of Q2 is at AC ground potential. Q1 is biased to cut-off and serves no active part and is disregarded. Notice the base electrodes are at AC ground potential, besides having a zero voltage base bias.

Fig. 7-19 shows the integrated package connected as an autodyne converter. In this configuration all the collector current of Q3 flows through Coil B to the voltage supply Vcc; hence, the current in the oscillator tuned circuit is independent of the received signal and is not subject to irregularities typical in other autodyne systems.

INTEGRATED PREAMPLIFIER PACKAGE

A commercial integrated package ideal for use as a low-level preamplifier is shown in Fig. 7-20. The IC portion consists of three transistors and four resistors and is designed for use as a general purpose preamplifier. The input may be a general purpose microphone, magnetic phono pickup, or tape head. The IC circuit provides so much gain in such a small space that the input and output leads must be oriented to prevent feedback oscillation.

Chapter 8

Transistor Oscillators

Basically an oscillator is an amplifier with regenerative feedback. Since we know that the transistor is capable of functioning as an amplifier, it must follow that by providing a positive feedback path from the output to the input, the unit will oscillate. A typical transistor audio oscillator is shown in Fig. 8-1. Another type of audio oscillator, resembling a Colpitts circuit, is shown in Fig. 8-2. Capacitors C1 and C2 form an audio tank circuit with a pair of earphones. A key connected in place of the "on-off" switch will make a satisfactory code-practice oscillator. Audio oscillators usually operate within a frequency range of 20 to 20,000 Hz, and by the selection of proper components, transistor oscillators may be designed for ultrasonic frequencies and on up to UHF.

When a particular frequency is desired, a circuit resonant at the desired frequency is used in the feedback path. It may be a tank circuit, a quartz crystal, or an RC network with a fixed time constant. Any of these systems may be used, provided that the feedback phase from the output to the input is in correct relationship to establish regeneration. Many modifications in basic oscillator circuit design are possible, and the type of circuit required will depend upon its ultimate use. A few circuit design features are listed below and illustrated in Fig. 8-3.

1. Impedance-tapped transformers for improved matching
2. Split capacitance across transformer windings for improved matching
3. Shunt-fed tuned circuits across resistive loads

4. Unbypassed emitter resistor to improve stability and reduce distortion.

EXTERNAL FEEDBACK OSCILLATORS

Several radio-frequency oscillator circuits are shown in Fig. 8-3. The stages illustrated use resonant circuits with

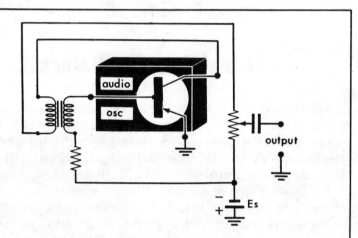

Fig. 8-1. Basic audio oscillator circuit using collector-to-base feedback. Notice the grounded-emitter configuration.

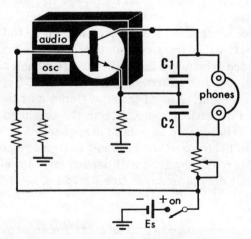

Fig. 8-2. This oscillator requires no feedback transformer. High-resistance magnetic earphones are connected so that the earpiece coils form a tank circuit with C1 and C2.

external feedback coupling provided by one of the following methods:

- Collector-to-base
- Collector-to-emitter
- Emitter-to-base

Diagrams 1 to 8 in Fig. 8-3 use PNP transistors; however, these stages will operate satisfactorily using NPN types.

Diagram 1: A resonant circuit is used in the high-impedance collector circuit to feed back an in-phase signal (positive feedback) to the base pickup coil. The base coil is designed and coupled to efficiently drive the moderately high-input impedance of the base circuit.

Diagram 2: In this circuit the feedback path is between the collector and emitter. Here the turns stepdown is greater since the emitter impedance is relatively low compared to the collector impedance.

Diagram 3: In this stage the feedback path is between the collector and the base, similar to the circuit in Diagram 1 with one exception: the resonant circuit is shunt-fed through capacitor C1. Since C1 is in series with the interelectrode capacity it provides a wider tuning range.

Diagram 4: This circuit employs a series LC circuit, similar to a shunt-fed circuit shown in Diagram 3; the difference being that the coupling capacitor is now an active impedance in the resonant tank circuit. This circuit bears some similarity to the Clapp oscillator. The LC series circuit may be replaced with a quartz crystal for better frequency stability.

Diagram 5: In this circuit the feedback path between the collector and emitter resembles the Colpitts circuit. The impedance ratio of C2 to C1 provides the stepdown impedance for driving the emitter input circuit. The impedance ratio of the two capacitors should be equal to the impedance ratio between the output and input circuits. C2 is usually 10 times greater than C1 but may not be a precise match. Capacitor C1 determines the frequency of the tuned circuit.

Diagram 6: Here the feedback path is connected between the collector and emitter and is somewhat similar to the circuit in Diagram 5, with one exception: The impedance stepdown is accomplished by a tap on feedback coil through a

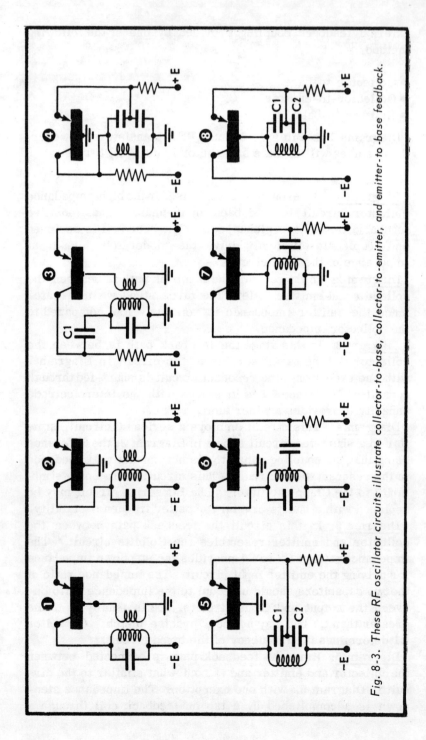

Fig. 8-3. These RF oscillator circuits illustrate collector-to-base, collector-to-emitter, and emitter-to-base feedback.

feedback capacitor. The inductance is similar to an auto-transformer, with a tap at the correct number of turns to establish an efficient match between the output and input circuits. Diagrams 2, 4, 5, and 6 are referred to as "common-base" oscillators and are usually selected for their stable performance.

Diagram 7: In this stage the feedback path connects the moderately-high-impedance base to the very-low-impedance emitter with circuitry similar to that in Diagram 6.

Diagram 8: This circuit employs another emitter-to-base feedback path with circuitry similar to that in Diagram 5.

The biasing used in the circuits illustrated in Fig. 8-3 requires a 3-terminal battery as shown in Fig. 8-4. An oscillator stage using a 2-terminal battery, self-bias, and stabilization requires bias resistors as shown in Fig. 8-5. In vacuum tube oscillators self-bias is permitted to build up due to grid current flow and the RC time constant of the coupling capacitor and grid resistor. A similar method of self-bias may be obtained in most transistor oscillator circuits. A self-biasing oscillator circuit is shown in Fig. 8-5. In this circuit the emitter current that flows during positive half cycles of emitter voltage charges capacitor C1 and builds up a Class C bias across R1. This stabilizes the output similar to a self-biased vacuum tube oscillator. To prevent blocking (squegging) the product of R1 and C1 must be small. R4 in the emitter circuit provides additional stability and permits the circuit to operate with a wider range of transistor constants.

CRYSTAL-CONTROLLED OSCILLATORS

Crystal control may be applied readily to transistor oscillators. The crystal may be used in conjunction with a resonant circuit. See Fig. 8-6. The crystal oscillators shown in Figs. 8-7 and 8-8 both use external feedback, whereas oscillators that depend upon feedback within the transistor are also considered. However, it is possible to incorporate crystal control in any of the oscillator circuits shown in Fig. 8-3. Also, frequency multiplication is possible by tuning the oscillator circuit to a harmonic of the crystal.

Frequency stability is dependent upon the crystal itself, while amplitude stability is dependent upon the characteristics

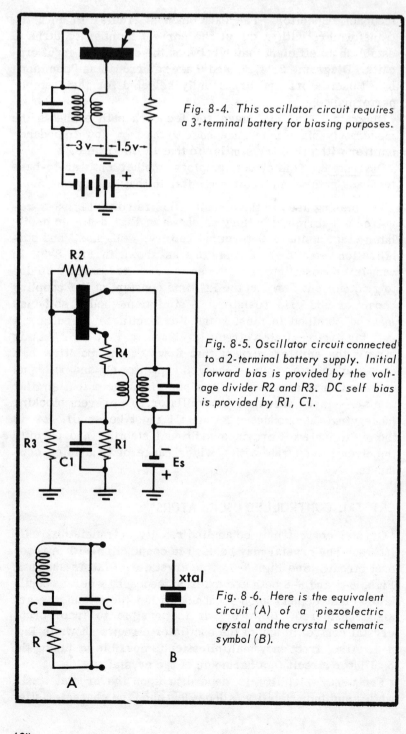

Fig. 8-4. This oscillator circuit requires a 3-terminal battery for biasing purposes.

Fig. 8-5. Oscillator circuit connected to a 2-terminal battery supply. Initial forward bias is provided by the voltage divider R2 and R3. DC self bias is provided by R1, C1.

Fig. 8-6. Here is the equivalent circuit (A) of a piezoelectric crystal and the crystal schematic symbol (B).

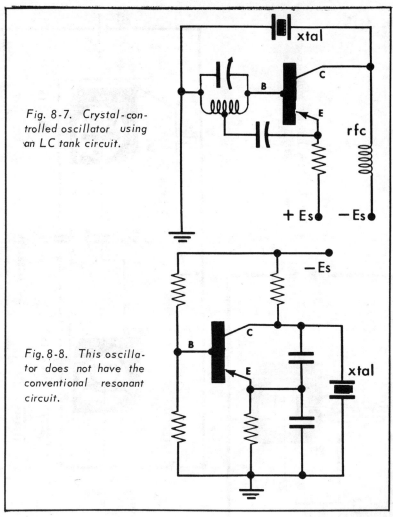

Fig. 8-7. Crystal-controlled oscillator using an LC tank circuit.

Fig. 8-8. This oscillator does not have the conventional resonant circuit.

of the transistor and its voltage supply. A self-biasing oscillator circuit similar to that shown in Fig. 8-5 would be desirable in maintaining constant amplitude and constant frequency if crystal-controlled.

MULTIVIBRATORS

It is possible to construct multivibrator circuits using transistors. Fig. 8-9 shows a vacuum tube plate-coupled multivibrator, and its transistor equivalent appears in Fig. 8-10. A transistor multivibrator circuit oscillates due to feedback

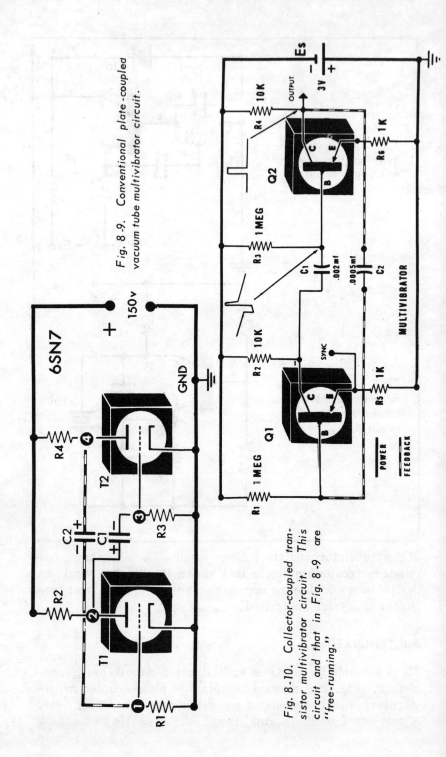

Fig. 8-9. Conventional plate-coupled vacuum tube multivibrator circuit.

Fig. 8-10. Collector-coupled transistor multivibrator circuit. This circuit and that in Fig. 8-9 are "free-running."

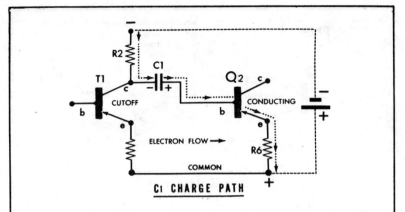

Fig. 8-11. When Q1 is cut off C1 charges through R6 as the collector voltage rises to maximum. The arrows indicate current flow.

from the collector of Q2 to the base of Q1; then from the collector of Q1 to the base of Q2 in a regenerative cycle.

Cut-off of a PNP transistor is achieved when the base is driven positive with respect to the emitter by the discharge of the coupling capacitor through the base resistor. The period of cut-off of each transistor is determined by the time constant of the discharge path of its respective RC coupling. The alternate charge and discharge paths of the coupling capacitors C1 and C2 are shown in Figs. 8-11, 8-12, 8-13, and 8-14.

BASIC OPERATION

When Q1 is cut off its collector voltage rises to maximum and charges coupling capacitor C1 through the conducting transistor Q2 and its emitter resistor R6. The arrows indicate the direction of electron current flowing through the charge path. See Fig. 8-11. When Q1 conducts, its collector voltage decreases and C1 discharges through R3, the conducting transistor Q1, and its emitter resistor R5. See Fig. 8-12. The voltage developed across base resistor R3 (Q2), due to the discharge of C1, establishes a reverse base bias, resulting in the cut-off of Q2. The cut-off period depends upon the time constant of the discharge path of C1.

When Q2 is cut off its collector voltage increases to maximum and charges coupling capacitor C2 through R4, the con-

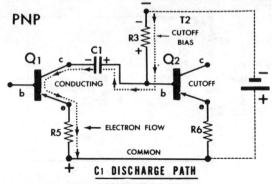

Fig. 8-12. As Q1 conducts its collector voltage decreases and C1 discharges through R3, Q1, and R5.

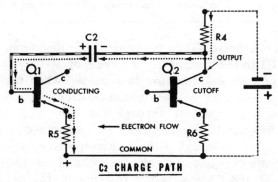

Fig. 8-13. C2 charges as Q2's collector voltage rises at cutoff.

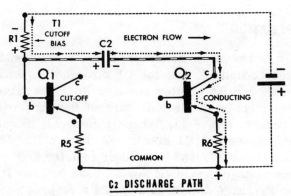

Fig. 8-14. When Q2 comes out of cut off, its collector decreases and discharges C2 through R1, Q2, and R6. The voltage developed across R1 (during the C2 discharges) establishes a reverse base bias, cutting off Q1.

ducting transistor Q1, and its emitter resistor R5. See Fig. 8-13. When Q2 comes out of cut-off, its collector voltage decreases and discharges C2 through R1, the conducting transistor Q2, and its emitter resistor R6. See Fig. 8-14. The voltage developed across the base resistor R1, due to the discharge of C2, establishes a reverse base bias, resulting in the cut-off of Q1. The cut-off period depends upon the time constant of the discharge path of C2.

When the collector and base resistors and coupling capacitors are equal (R2 = R4, R1 = R3, C1 = C2), and the transistors have comparable characteristics, the output signal will be symmetrical. Any deviation from this balance will produce a non-symmetrical output, the degree of which will depend upon the difference in circuit component values. These may be varied to change the frequency and symmetry of the output signal.

EMITTER-COUPLED MULTIVIBRATOR

An emitter-coupled multivibrator is shown in Fig. 8-15. The operation of this circuit is illustrated by four oscillograms photographed during tests. The charge and discharge paths are given to coincide with the direction of electron flow; i.e., negative to positive.

Operation:

1. When the voltage supply (Es) is applied, transistor Q2 conducts and capacitor C1 charges through R2, Q2 (base to emitter) and, finally, R5.

2. Notice the emitter current of Q2 and the charge current of C1 both flow through R5 (2.2K). This resistor is common to both emitters.

3. The voltage across R5 develops a negative potential at the emitter of Q1 (point 3) and is of sufficient magnitude to cut off this transistor.

4. As C1 nears full charge the voltage across R5 decreases and allows Q1 to conduct.

5. When Q1 conducts it causes C1 to discharge collector to emitter and R3 (330K).

6. The discharge current develops a voltage across R3 that is positive at the base of Q2, causing this transistor to cut off.

7. R3, being a relatively large resistance (330K), makes the time constant (R3-C1) relatively long; therefore, Q2 will be cut-off for a longer period than Q1.
8. When C1 has discharged sufficiently, Q2 conducts, causing C1 to recharge and the cycle starts over again.

Study carefully the charge and discharge paths of C1 and compare the RC time constants. (Assume the resistance through each transistor to be negligible when conducting.)

ECCLES-JORDAN MULTIVIBRATOR (Vacuum Comparisons)

Fig. 8-16 is an Eccles-Jordan trigger circuit using vacuum tubes. This type of multivibtator employs direct coupling

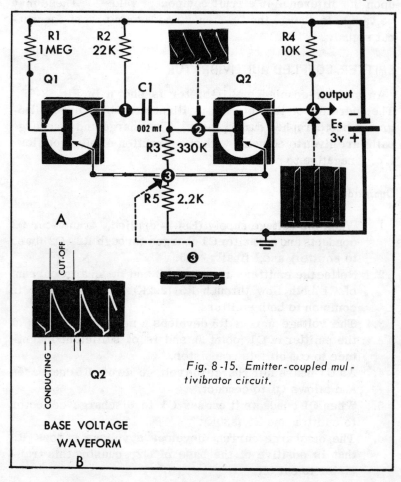

Fig. 8-15. Emitter-coupled multivibrator circuit.

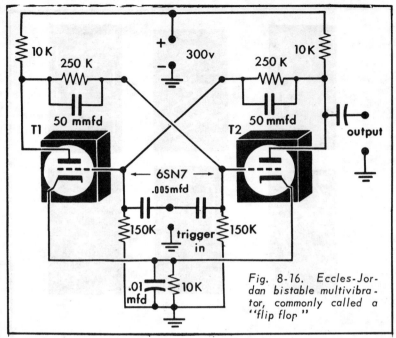

Fig. 8-16. Eccles-Jordan bistable multivibrator, commonly called a "flip flop"

(DC) between the plates and grids of both tubes.   The circuit is designed to have two states of stable equilibrium, they are:

1.   T1 conducting, T2 cutoff
2.   T2 conducting, T1 cutoff

This system is said to flop from one stable state to another when triggered, and is generally referred to as a "flip-flop" circuit. The following steps explain the basic functions of the circuit:

1.   To facilitate the discussion, let us assume that the filaments of both tubes are heated, and when the "B" supply is applied T1 starts to conduct more plate current than T2. This will lower the plate voltage of T1 and at the same instant reduce the grid potential (less positive) of T2. Remember that a grid may be positive with respect to ground but negative with respect to cathode.
2.   This, in turn, reduces the plate current of T2, and increases its plate voltage. The increasing plate voltage of T2 drives the grid of T1 more positive, causing T1 to conduct heavily.

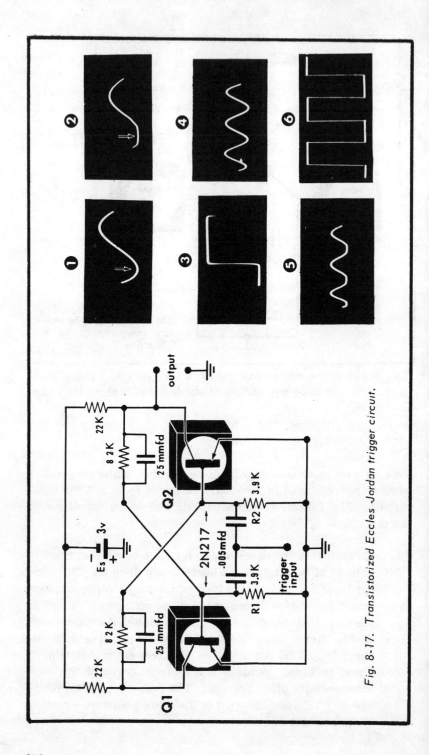

Fig. 8-17. Transistorized Eccles-Jordan trigger circuit.

3. This action is cumulative, until the grid voltage of T2 is sufficiently negative with respect to its cathode, causing T2 to cut off. When this condition is reached the circuit is stabilized, with T1 conducting heavily and T2 cut off. This stable condition will prevail until the circuit is triggered.

4. If a capacitively-coupled positive trigger pulse of sufficient magnitude is applied to both grids, it is evident that T1, the conducting tube, will be unaffected, while T2 will be driven into conduction. This causes an immediate drop in the plate voltage of T2, thereby lowering the grid voltage of T1.

5. This causes regeneration (feedback) to occur, but this time in the reverse direction, and the cumulative action causes the circuit to flop over to its second stable state of equilibrium.

6. When this condition is reached the circuit is again stabilized, where T1 is cut off and T2 is conducting heavily. The circuit will remain at rest (quiescent) until triggered.

Fig. 8-17 shows the transistorized version of the Eccles-Jordan circuit. In this case it is current feedback that causes regeneration when triggered. The oscillograms illustrate the operation of the transistorized circuit. When a sine wave of voltage is applied to the base of Q1, alternate cut-off and conduction switches back and forth between Q1 and Q2. This provides a square-wave output at the collector of Q2. While the sine wave is positive-going, Q1 is cut off; it conducts when the sine-wave swing is negative-going; therefore each sine wave produces a square-wave signal, the frequency of the output being equal to the frequency of the input.

Oscillogram 1 shows a single sine wave of the input signal before it is connected to the circuit. Oscillogram 2 shows the same sine wave when connected to the input of the multivibrator. Notice the distortion of the negative peak. This is due to the conduction of Q1, which represents a low-impedance load, similar to a conducting diode. Oscillogram 3 shows the square-wave output as it appears at the collector of Q2. Oscillograms 4, 5, and 6 show the same waveforms using a 3-cycle pattern. Notice, it is the negative-going signal voltage that causes Q1, a PNP transistor, to conduct. However, using an NPN transistor the reverse will be true.

SCHMITT TRIGGER CIRCUIT

The circuit diagram shown in Fig. 8-18 is known as the Schmitt trigger circuit, which consists of a pair of pentode tubes coupled as a multivibrator. Besides the direct coupling between T1 and T2, the common cathode resistor provides cathode-follower feedback coupling. The operation of the circuit is as follows:

1. Let us assume that the input signal (trigger input) to T1 is a steady voltage that can be varied.
2. When this voltage is zero, T1 is cut off, while T2 is conducting heavily. The plate current of T2 passing through the common cathode resistor provides sufficient bias to hold T1 at cut off. Hence, T2 is operating as a cathode follower.
3. If the input signal voltage (trigger) to T1 is raised positively, nothing happens until it reaches a value nearly equal to the common-cathode voltage.

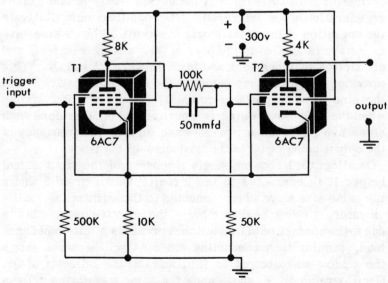

Fig. 8 -18. Basic vacuum tube Schmitt trigger circuit.

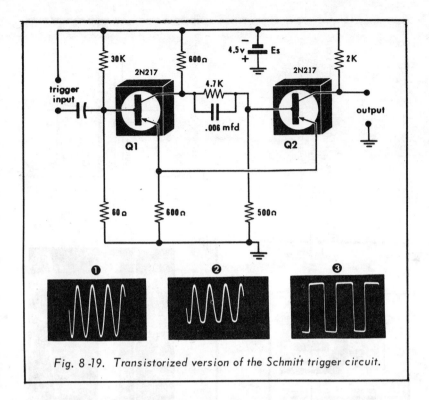

Fig. 8-19. Transistorized version of the Schmitt trigger circuit.

4. When this condition occurs T1 will conduct heavily, causing the grid bias on T2 to become less positive with respect to the cathode. Also, the plate current of T1 flowing through the common cathode resistor develops a voltage which is additive, causing T2 to cut off. Thus the plate voltage of T2 rises sharply from minimum to maximum.

5. The 50-pfd capacitor is used to produce a fast transition from one state to another.

6. When the positive input voltage to T1 is reduced or removed the circuit returns to its original state; that is, T1 cut off and T2 conducting. Trigger action may be accomplished by applying a sine-wave signal. In this case T1 will conduct during the positive-going swing and cut off during the negative swing.

The circuit in Fig. 8-19 is the transitorized version of the Schmitt trigger circuit. Oscillogram 1 shows the input signal before it is connected to the circuit. Oscillogram 2 shows

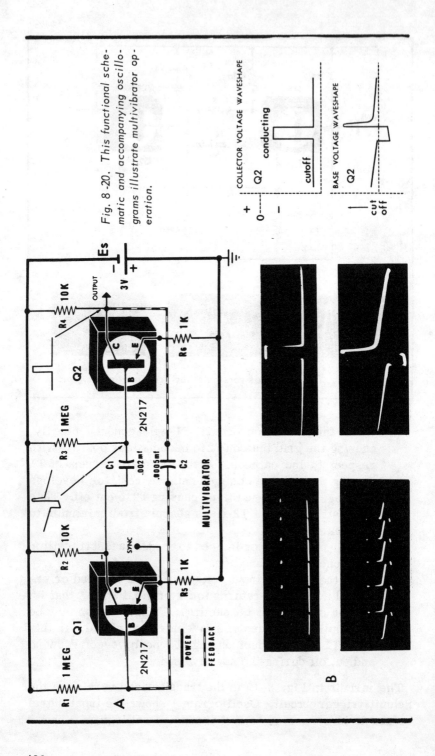

Fig. 8-20. This functional schematic and accompanying oscillograms illustrate multivibrator operation.

the same signal when connected. Notice the distortion of the negative peaks. Oscillogram 3 shows the square-wave output during input trigger action.

One purpose of this study is to enable the reader to observe the base and collector waveforms with an oscilloscope, as well as to study the charge and discharge of the coupling capacitors. Actual oscillograms taken at the base and collector terminals of Q2 are shown in Fig. 8-20. The schematic illustrates circuit operation employing the following component values:

All resistors:	1/4 watt carbon
Capacitors:	C1 = .002 mfd
	C2 = .0005 mfd
Power Supply:	3-volt battery
Frequency:	Approximately 270 Hz

The schematic in Fig. 8-21 is a complementary-symmetry (PNP-NPN) multivibrator. Since the circuit is obviously unique to transistor (no vacuum tube equivalent) its operation is interesting:

1. When power is applied both transistors start to conduct and charge capacitor C1. The electron current flows from the negative terminal of the battery through Q2 (emitter-collector) to the negative side of C1, and through the base-emitter junction of Q1 to the positive terminal of the battery.
2. The charge current flowing through the Q1 base-emitter junction increases the collector current, which flows through the base-emitter junction of Q2.
3. The relatively high Q2 base current causes a relatively heavy collector current to flow, which rapidly charges C1.
4. As the charge current of C1 decreases (exponentially) the base current of Q1 decreases; this in turn reduces the base current of Q2. This condition is cumulative until the base current of Q1 is near zero. At this moment C1 discharges.
5. C1 discharges through R1 and R2. The discharge current flowing through R1 develops a positive bias voltage at the base of Q1, causing it to cut off, which in turn drives Q2 into cut-off.

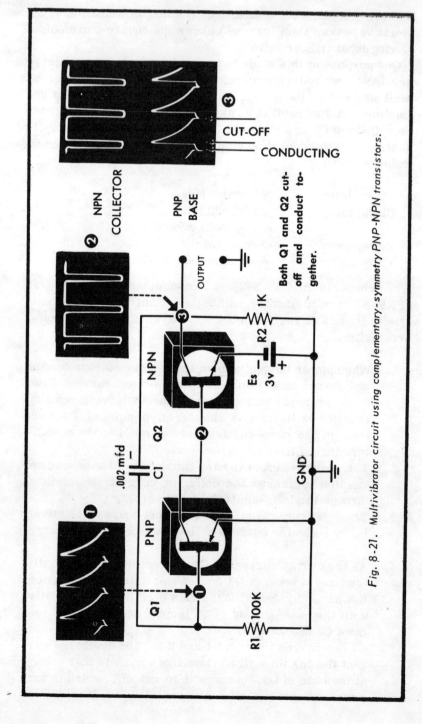

Fig. 8-21. Multivibrator circuit using complementary-symmetry PNP-NPN transistors.

Both Q1 and Q2 cut-off and conduct together.

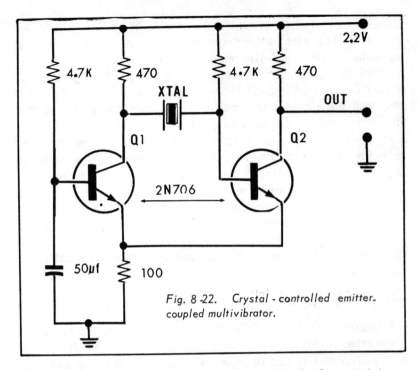

Fig. 8-22. Crystal - controlled emitter-coupled multivibrator.

6. When C1 has discharged sufficiently, both transistors begin to conduct and the cycle of charge and discharge is repeated; the repetition rate depends upon the RC time constant of R1, R2, and C1 in the discharge path. Since the RC time constant of the charge path is much smaller, the output square wave at the Q2 collector is asymmetrical, as shown in Oscillogram 3. Notice the long discharge curve of C1 in Oscillogram 1.

CRYSTAL-CONTROLLED MULTIVIBRATOR

In order to develop an accurate pulse generator a piezo-electric crystal is used to replace the coupling capacitor in the emitter-coupled square-wave oscillator in Fig. 8-22. The circuit functions similar to a conventional multivibrator, except that the output of Q1, originally used to charge a timing capacitor, is now exciting a crystal which provides a stable time constant.

COLPITTS TRANSISTOR OSCILLATOR

The oscillator circuit in Fig. 8-23 requires no feedback

transformer and is similar to the circuit in Diagram 5 in Fig. 8-3, but with some circuit modifications to provide an audio output. Oscillogram 1 shows the waveform across a 2000-ohm resistor connected in place of the earphones, while Oscillogram 2 shows the waveform appearing at the output with the earphones connected. Notice the effect of the inductive load on the waveform (frequency 1100 Hz).

Fig. 8-24 is a junction transistor blocking oscillator similar in many respects to the vacuum tube blocking oscillator long popular in TV circuits. It is interesting to note that blocking oscillator transformers designed for vacuum tube circuits have been used successfully with transistors. With the resistance values given in the diagram, a frequency range of 15 to 1 is not uncommon. Care should be exercised to see that induced voltages do not exceed the transistor ratings, though.

COUNTING BY FLIP-FLOP

A typical bistable multivibrator using indicator lamps is illustrated in Fig. 8-25. The unit is generally referred to as a flip-flop and is one of many such units used in computers. Each stage in a bank operates as a switch and has two stable states—ON or OFF. In vacuum tube or transistor flip-flop stages, these stable states are usually referred to as "conducting" meaning ON, or "non-conducting" meaning OFF.

The two stages are interconnected in such a manner that

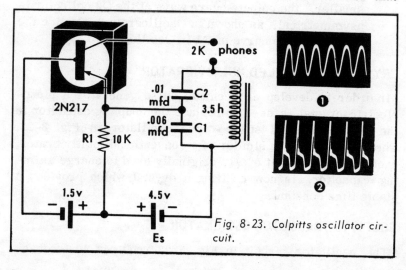

Fig. 8-23. Colpitts oscillator circuit.

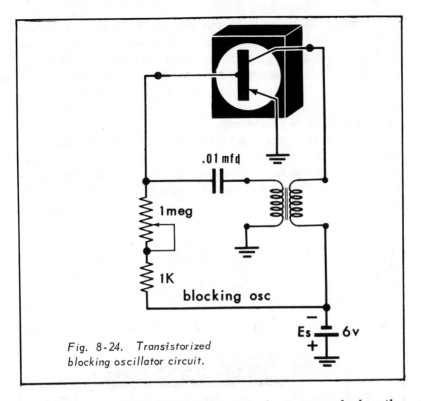

Fig. 8-24. Transistorized
blocking oscillator circuit.

when Q1 is conducting Q2 is non-conducting, and when the
circuit is triggered the opposite condition is true. A colored
lamp, connected in the collector circuit of each stage, lights
when the stage is conducting. For example, if Q1 is conducting the light will be GREEN and when Q2 conducts the
light will be RED. These colors are important since the total
count is made using the green lights in the chain.

To understand how a digital computer counts, it is necessary to study a series chain of flip-flop units which we will
call Unit 1, Unit 2, Unit 3, and so on. However, we will
study a single unit as shown in Fig. 8-25 and then add others
as we proceed with the study. When power is applied to the
first unit, one transistor will start conducting a little faster
than the other. This is due to minor differences in transistors and components. The one that has the head start will
hold the other at cut-off and a stable state is established
which will continue until triggered.

Let us assume that Q1 was the first to conduct; this will be
indicated by the green light, and, since this unit is the first

in the chain the green light will indicate a count of 1, which is an error since the counting procedure has not been started. To "made ready," we must extinguish the green light by applying a positive trigger pulse to the "reset" terminal. This triggers the circuit, causing Q1 to cut off, which in turn drives Q2 into conduction as shown by the red light and indicates a count of "0" (zero).

The unit is now ready to start counting, so let us begin with the count of 1 by applying one positive pulse to the count terminal. Remember, Q2 is conducting and the red lamp is illuminated. When a positive pulse is applied to its base, this stage (Q2) is cut off which, in turn, drives Q1 into conduction, turning on the green lamp and extinguishing the red. The panel indication now shows the green light "on" and the red lamp "off" which indicates a count of "one."

It is helpful at this point to study what happened to the Q2 collector voltage for a count of 1. When this stage (Q2) cut off, the collector voltage increased negatively (PNP); this

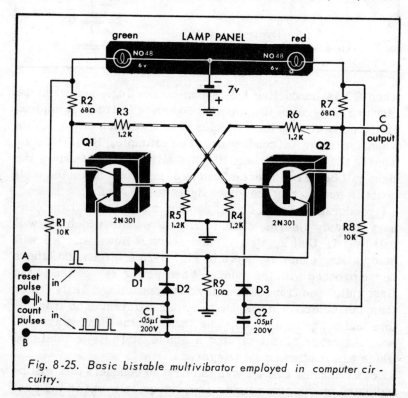

Fig. 8-25. Basic bistable multivibrator employed in computer circuitry.

constitutes a "negative pulse." Now, if this pulse is applied to the count input terminal of Unit 2 (this has been reset), it would have no effect since it requires a "positive pulse" to trigger. Hence, the count of one was not passed on to the second unit. A second positive pulse for a count of 2 to Unit 1 will cut off Q1 and drive Q2 into conduction, causing the green light to extinguish and the red light to illuminate. Now a count of 2 is the limit for the first unit.

Let us now consider Unit 2, which consists of a flip-flop— Q3 and Q4. See Fig. 8-26. Helpful here is a study of the Q2 collector voltage (first unit) when it was triggered for the count of two. Since Q2 conducted on the count of two, the collector voltage dropped from a relatively high negative voltage to a lower value (PNP). This drop constitutes a positive-going pulse at the collector (Q2). This, in turn, triggered Unit 2, causing Q4 to cut off and Q3 to conduct, illuminating the green lamp, indicating a count of 2. Notice that the count of 2 cleared Unit 1 and is indicated on Unit 2 by the green lamp. (Q3 and Q4 of Unit 2 now shown.) Notice that two positive pulses applied to the input of any unit provides only one positive pulse at the output. Therefore, a green light on a third unit would indicate a count of 4; a green light on a fourth unit would be a count of 8; a fifth unit, a count of 16 and so on.

BINARY COUNTING

The arrangement of the red and green panel lamps for six flip-flop units is illustrated in Fig. 8-26. In order to correspond with the sequence of binary numbers, the lamps (green and red) of Unit 1 are located on the right side of the panel, since they represent the first division of count 2. Notice the green lamp is mounted above the red lamp. The lamps of Unit 2 are positioned in the same manner color-wise but this pair is located to the left of Unit 1. As each unit is added their respective panel lamps are arranged in the same order colorwise but located progressively to the left on the panel. This provides a row of green lamps (top) and a row of red lamps (bottom), thus making it convenient to read.

The green lamp furthest to the left represents the highest count in powers of 2 for the six flip-flop units illustrated. For example, if 32 positive pulses are applied to the count input of Unit 1 (after all units have been reset) the green lamp

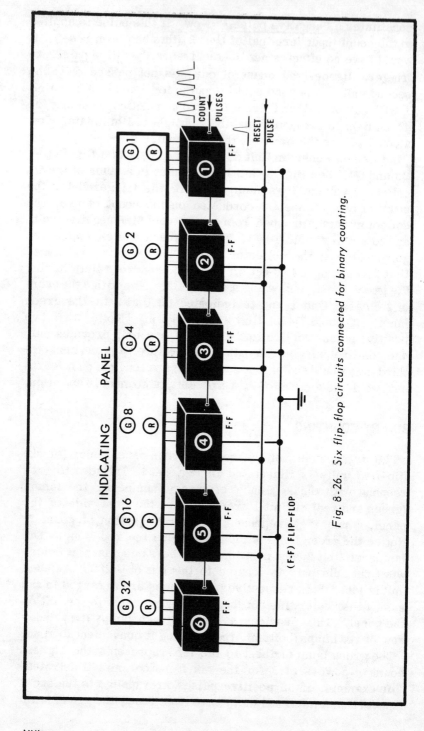

Fig. 8-26. Six flip-flop circuits connected for binary counting.

Decimal Number	Binary Number	Decimal Number	Binary Number
0	0	17	10001
1	1	18	10010
2	10	19	10111
3	11	20	10100
4	100	21	10101
5	101	22	10110
6	110	23	10111
7	111	24	11000
8	1000	25	11001
9	1001	26	11010
10	1010	27	11011
11	1011	28	11100
12	1100	29	11101
13	1101	30	11110
14	1110	31	11111
15	1111	32	100000
16	10000	•	•

of Unit 6 is illuminated, followed by five illuminated red lamps. Hence, we read out a binary number of 100,000 for six units operating—Green "ON," count 1; Red "ON," count 0—as we read from left to right; therefore, a count of 32 equals one green followed by five reds.

Let us assume that you are making the count of 32. When you arrive at 31 you will notice, reading from left to right, a red lamp illuminated followed by 5 green illuminations; therefore, you would record the following binary count: 011111. This binary number will represent a decimal count of:

color: R G G G G G

indications: 0 + 16 + 8 + 4 + 2 + 1 = 31

A table giving decimal numbers from 0 to 32 and their binary equivalents appears in the accompanying chart. The six flip-flop units limit the binary count to 111111 (all greens on) and when converted to decimals is: 111111 = 32 + 16 + 8 + 4 + 2 + 1 = 63.

Each green light, when illuminated, represents a power of 2 and is recorded in binary numbers as 1 and, if extinguished, as 0. Although we have only "ones" and "zeros" to record, it is their position from left to right that determines their power of 2. The first green lamp we read in the diagram is 2^5 which is 32; the second from the left is 2^4 which is 14; the third from the left is 2^3 which is 8; and so on. (See the Table.) To read out the total, use the green lights that are illuminated and add up their respective digits. For example, if every other green lamp is illuminated in the 6-unit flip-flop, the binary count will be: 101010 = 32 + 8 + 2 = 42. The six flip-flops in the demonstration are used to explain the binary number system: for higher counts we must connect additional flip-flops.

Chapter 9

Power Supplies

The power transistor has made possible a new system of voltage conversion—an improved method for obtaining a relatively high-voltage DC from a low-voltage battery. For example, many mobile electronic devices require high-voltage DC and are dependent upon vibrator or dynamotor systems. Such mechanical high-voltage DC sources are bulky and require frequent inspection and overhaul. Conversely, the transistorized power supply has no moving parts; it is compact, and has an efficiency rating as high as 90%. With these physical advantages and such exceptional performance, the transistorized power supply is rapidly replacing vibrator and dynamotor systems.

OSCILLATOR-TYPE POWER SUPPLY

A simple oscillator-type power supply using two power transistors is shown in Fig. 9-1. This circuit operates on a 1.5-volt battery (Es) and will deliver a considerably higher output voltage. The extent of this increase is governed by the specified wattage rating of the transistors when used with suitable heat sinks.

The base drive for transistor Q1 is obtained from the collector terminal of Q2 and vice versa. This crossover circuit provides the correct feedback phase between Q1 and Q2 to sustain oscillation. The base current is controlled by a series limiting resistor connected in the base lead of each transistor. This resistance has a permissible minimum of 10 ohms for a 1.5-volt input (Es) and is increased when higher input voltages are used.

SEPARATE BASE DRIVE

The operating efficiency of the circuit is improved by using a separate feedback winding for base current drive as illustrated in Fig. 9-2. This winding is center-tapped and has a stepdown ratio of 20 to 1. Base resistor R2 (20 ohms) forms a voltage divider with rheostat R1. This control is used to adjust the base current of both transistors for normal operation. Transformer T1 has a 3-watt rating which is capable of delivering 250 volts (DC) at 10 milliamperes for an input voltage (Es) of 6 volts. The output secondary has a stepup ratio of 50 to 1.

Operation:

1. On closing the power switch (s) collector current starts to flow in one transistor, since minor differences exist between transistors of identical types.
2. Let us assume that the Q1 collector current has a head start and rises sharply. This current flowing through the top half of the primary induces a voltage across the base winding; the voltage is phased to provide a FORWARD BIAS to the base of Q1, which conducts heavily. The other end of the base winding applies a REVERSE BIAS to the base of Q2 and drives this transistor to cut-off.

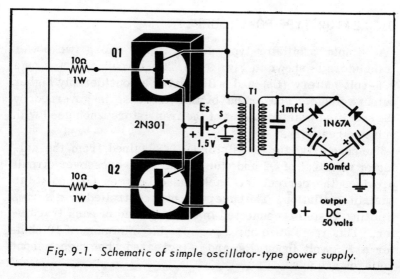

Fig. 9-1. Schematic of simple oscillator-type power supply.

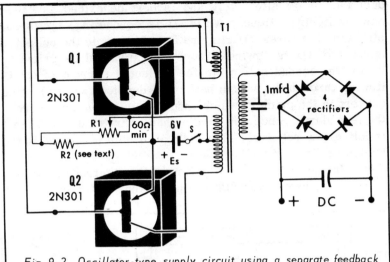

Fig. 9-2. Oscillator-type supply circuit using a separate feedback winding for base current drive.

3. The POSITIVE* feedback to the base of Q1 and the NEGATIVE feedback to the base of Q2 continues until the Q1 collector current reaches saturation.

4. At this moment the induced voltage across the base winding ceases to exist.

5. When this occurs the Q1 collector current falls rapidly and the collapsing magnetic field induces a voltage of opposite polarity across the base winding.

6. The reversed base bias voltage applies a FORWARD BIAS to the base of Q2 (previously cut off) and a REVERSE BIAS to the base of Q1 (previously conducting).

7. At this moment the Q2 collector current rises sharply through the bottom half of the primary, and Q1 is cut-off.

8. This condition continues until the Q2 collector current reaches saturation and the induced voltage across the base winding ceases to exist.

9. At this moment the Q2 collector current collapses, inducing a voltage of opposite phase across the base winding, and the cycle of events is repeated.

The regenerative cycle causes the two transistors to con-

---

*POSITIVE feedback sustains oscillations—NEGATIVE feedback suspends oscillations.

duct and cut off alternately and is best described as "see-saw" switching. Hence, while Q1 is conducting, Q2 is cut off, and vice versa. During the first half cycle the collector current of Q1 is flowing through the top half of the primary; during the second half cycle the collector current of Q2 is flowing through the bottom half. This develops an alternating current in the primary which induces an AC voltage across the secondary. Since this winding has a turns ratio of 50 to 1 it delivers a relatively high voltage, which is applied to the bridge rectifier circuit.

The "see-saw" switching action of the two transistors develops a square-wave voltage output. Since the collector load

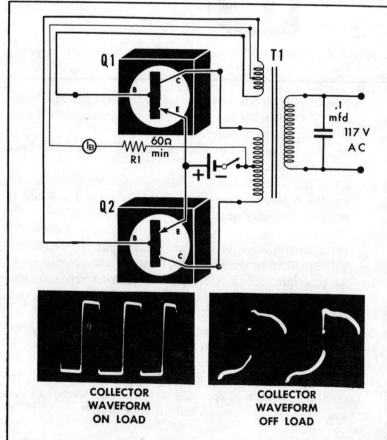

COLLECTOR
WAVEFORM
ON LOAD

COLLECTOR
WAVEFORM
OFF LOAD

Fig. 9-3. The oscillograms show typical waveforms encountered in this AC power supply. Notice the 1-mfd filter across the secondary.

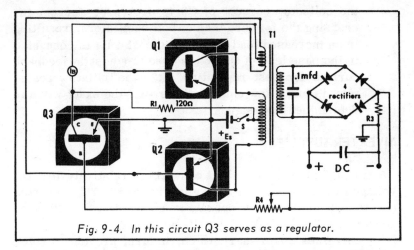

Fig. 9-4. In this circuit Q3 serves as a regulator.

is inductive, a .1-mfd capacitor is connected across the
secondary to filter out the transient spikes developed by the
leading and trailing edges of the waveform. This capacitor
also protects the transistors from high-voltage surges and
eliminates spurious frequencies. The repetition rate of the
square wave is about 60 Hz; therefore, with an input voltage
of 3 volts and a correctly adjusted base current the secondary
will deliver about 117 volts AC and may be used to power 60-
Hz equipment. See Fig. 9-3. In either case, AC or DC loads
that exceed the capability of the source will cause the system
to stop oscillating; therefore, overloads or short circuits
will not cause damage. By using a third transistor the output
voltage can be regulated as illustrated in Fig. 9-4.

Operation:

1. An increase in load current causes the "top" of R3 to
   become more negative.
2. This increases the FORWARD BIAS applied to regulat-
   ing transistor Q3, causing the collector current of Q3
   to increase through R1 and oppose the base bias of Q1
   and Q2.
3. This, in turn, decreases the collector current in both
   oscillators, thus counteracting an increase of load
   current.
4. When the load current through R3 decreases, regulat-
   ing transistor Q3 receives less forward bias, causing

the collector current of Q3 to decrease, thereby increasing the forward bias on both oscillators, resulting in an increased load current. Rheostat R4 is connected in the base lead of Q3 and is used to adjust the feedback current for best regulation. R3, the initiating resistor, is just a few ohms. However, the exact value is determined in design.

INVERTER SUPPLIES

A transistorized system that converts a relatively low DC voltage to 117 volts AC is shown in Fig. 9-5. This inverter (DC to AC) may be used to operate small electrical appliances requiring 117 volts AC. The self-sustaining push-pull oscillator energizes the primary which, in turn, provides a voltage stepup across the output secondary. The output is a square wave of voltage at approximately 60 Hz. Capacitor C1 removes spurious frequencies.

Notice that the pilot light filament is in series with the base input resistance, functioning as an automatic control on the base current. For example, to start oscillation a relatively large base current is necessary. The resistance of the filament when cold is very small, thus permitting a large base

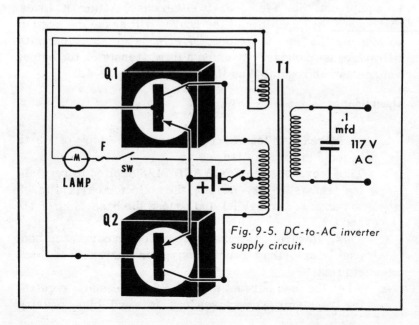

Fig. 9-5. DC-to-AC inverter supply circuit.

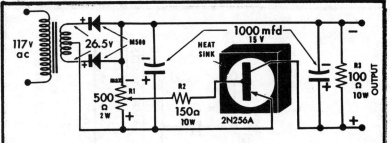

Fig. 9-6. Transistorized fullwave power circuit *supply* with an output variable from 0 to 12 volts DC. The transistor operates as a variable resistance in series with the output resistance R3. The input ripple at the base electrode is 180° out of phase with the ripple appearing at the collector, resulting in increased filtering action. R1 is the control potentiometer and R2 is the base current limiting resistor. Current output up to .5 ampere may be obtained.

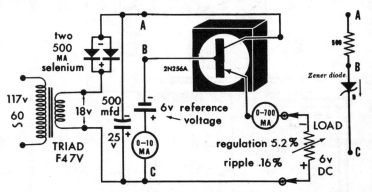

Fig. 9-7. Regulated power supply circuit in which the output voltage remains constant for wide variations of input (value of secondary voltage should never exceed 25 RMS). The output voltage is less than one half volt higher than that of the reference battery. Different reference voltages may be used for other output voltages. 6v reference battery may be replaced by a Zener diode. (National type A5B or equivalent.)

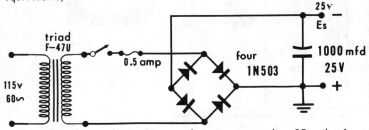

Fig. 9-8. This simple bridge rectifier circuit supplies 25 volts from 117 volt line.

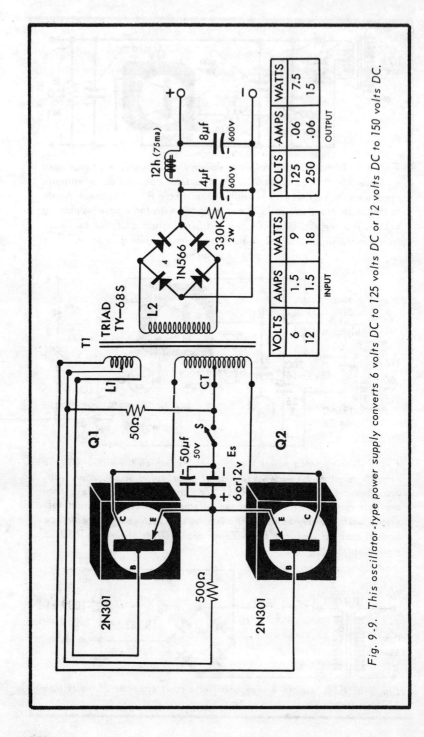

INPUT

VOLTS	AMPS	WATTS
6	1.5	9
12	1.5	18

OUTPUT

VOLTS	AMPS	WATTS
125	.06	7.5
250	.06	15

Fig. 9-9. This oscillator-type power supply converts 6 volts DC to 125 volts DC or 12 volts DC to 150 volts DC.

current to flow to start oscillation; however, as the filament heats up its resistance increases and lowers the base current to its normal operating level. Should the load short circuit, the inverter ceases to function, thus preventing damage to the components. It must be remembered that the collector electrode of the power transistor is common to its mounting flange; therefore, the case of the transistor must be insulated from the chassis (heat sink). Figs. 9-6, 9-7, and 9-8 are AC line-operated low-voltage DC supplies.

COMMERCIAL POWER SUPPLY

A transistorized oscillator-type power supply that converts 6 volts DC to 125 volts DC, or 12 volts DC to 150 volts DC is illustrated in Fig. 9-9.

Operation:

When power switch S is closed, collector current starts to flow in one transistor, inducing a voltage across the base coil. This induced voltage is phased to hold the other transistor at cut-off and to drive the conducting transistor into saturation. For example, let us assume that Q1 is the first to conduct. Its collector current flowing through the top half of the primary induces a voltage across base coil. L1.

The base coil voltage is phased so that it is negative at the base of Q1 and positive at the base of Q2. Since both transistors are PNP types, Q1 is driven into saturation due to positive feedback which provides an accelerating forward bias, while Q2 is cut off due to a reverse bias caused by negative feedback. Notice that it is the feedback from Q1 that causes Q2 to cut off. However, when Q1 saturates, induction between the top half of the primary and the base coil ceases to exist, causing the collector current of Q1 to collapse from saturation. This induces a voltage of opposite polarity across the base coil, resulting in sharp Q1 cut-off and a sharp rise of Q2 collector current through the bottom half of the primary. This condition will continue until Q2 reaches saturation. At this moment the feedback to both transistors ceases to exist, causing the Q2 collector current to collapse rapidly.

HIGH-VOLTAGE POWER SUPPLY

Fig. 9-10 is a transistorized high-voltage power supply designed for a 5 KV output with 26 volts input.

Operation:

1. A single-stage sine-wave oscillator, using collector-to-base feedback, operates into a 300-turn center-tapped primary coupled to a 20,000-ohm secondary, and is followed by a voltage doubler using two 1N3286 silicon rectifiers. These diodes have a peak inverse voltage rating of better than 3000v and a low leakage rating.

2. The output is filtered by two .01-mfd high-voltage capacitors which are active components of the doubler circuit. Additional filtering may be used to reduce the AC ripple. Operational efficiency is about 22%.

3. The transformer core is the same type used in TV flyback transformers. The primary winding is No. 38 Mylar wire totaling approximately 40 ohms (total) DC resistance.

EMERGENCY FLUORESCENT LAMP

A transistorized portable fluorescent lamp circuit is illustrated in Fig. 9-11. The system is similar to the transistor oscillator power supply previously discussed in which the feedback occurs between the collector-to-base circuits through N1 and N2, respectively. Resistors are connected in series between the negative terminal of the voltage supply (Es) and the base coil center tap. A 100-ohm potentiometer is connected as a rheostat so that the base current of both transistors may be controlled for normal operation. The oscillator

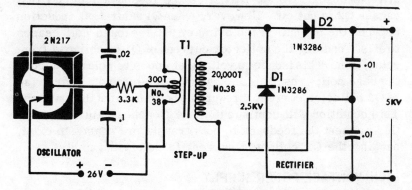

Fig. 9-10. This circuit is capable of a 5 KV output from a 26-volt battery.

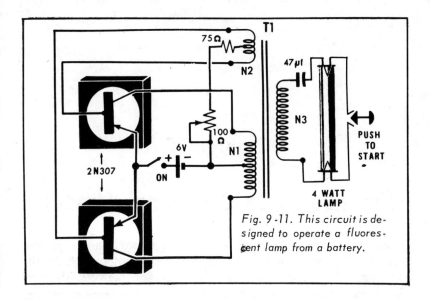

Fig. 9-11. This circuit is de-signed to operate a fluores-cent lamp from a battery.

requires extra base current to start and less to sustain os-cillations.

Since fluorescent lamps require a relatively high voltage for starting and operation, a stepup transformer (Triad Ty-100) is used to couple the oscillator output (N2) to the lamp through secondary N3. The frequency of the supply voltage is approximately 5 kHz; therefore, a capacitor is connected in series with the secondary (N3) and the lamp, functioning as a current-limiting ballast. The primary coil (N1) is 50 turns of No. 18 enameled wire; N2 is 50 turns of No. 30 enam-eled wire; both coils (N1, N2) are interlaced. The secondary (N3) is 600 turns of No. 30 enameled wire.

The system may be powered with a 5- or 6-volt battery. The unit was tested with a 5-volt sintered-plate nickel-cad-mium battery and supplies 1 ampere for two hours. This battery is, of course, rechargeable. The unit also operates satisfactorily on a 6-volt car battery. Should the load ter-minals short circuit, the oscillator ceases to function, thus preventing damage to the components.

Chapter 10

The Reactance Diode or Varactor

When P-type and N-type semiconductor materials are joined to form a diode, the junction becomes a "potential hill" or barrier. This develops a depletion region between the two electrodes where no mobile carriers exist, thereby making a "capacitor" which exhibits features not present in a regular capacitor. For example, reverse biasing a junction diode increases the width of this depletion region in direct proportion to the applied bias voltage. It provides automatic control of the dielectric width and is, in effect, a voltage-variable capacitor where frequency is a function of voltage. Such a device is referred to as a "reactance diode" or "varactor."

FREQUENCY MULTIPLIER

The reactance diode has many useful applications such as automatic frequency control and frequency multiplication. Theoretically, the device introduces distortion in a sine wave of current that is equal to all multiples of the fundamental frequency without the loss of power. This is ideal for harmonic generators where it is desired to obtain power at a multiple of a given fundamental. The P-N junction and symbol are shown in Fig. 10-1. When a sine wave is applied to the varactor the harmonic distortion resulting from the nonlinear variation of the capacitance to the sine-wave voltage makes it ideal also for frequency conversion. By connecting it in common to two resonant circuits, one tuned to the fundamental and the second tuned to twice the fundamental, the varactor becomes the heart of a simple frequency-doubler circuit. See Fig. 10-2.

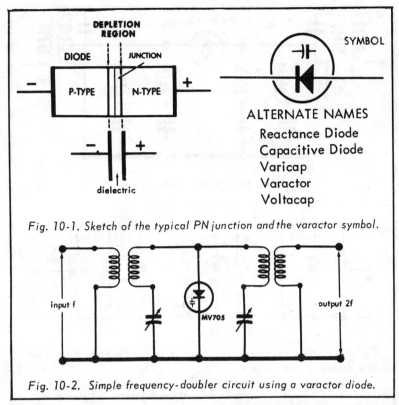

Fig. 10-1. Sketch of the typical PN junction and the varactor symbol.

Fig. 10-2. Simple frequency-doubler circuit using a varactor diode.

Practically any nonlinear device that distorts a sine wave can be used for frequency multiplication—a Class C amplifier, for instance. However, the reactance diode is more suited for frequency multiplication in the range of 50 to 1000 MHz and higher. A frequency quadrupler circuit using a varactor diode is shown in Fig. 10-3. The series resonant circuits ground out the 2nd and 3rd harmonics and produce a good sine wave of the 4th harmonic at the output.

The frequency multiplier circuit illustrated in Fig. 10-4 uses a varactor diode MV705. The input and output circuits are both double-tuned. The input (50 MHz) consists of tuned circuits L1-C1 and L2-C3, coupled by C2. The output (200 MHz) is composed of tuned circuits L4-C5 and L5-C7, coupled by C6. Coil L3 and capacitor C4 constitute an idler circuit used for frequency quadrupling but not for frequency doubling. A large resistance (R) is shunted across the varactor diode to provide self bias. The value is determined by the desired frequency multiplication factor; a typical value is 100K. The

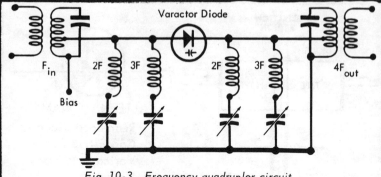

Fig. 10-3. Frequency quadrupler circuit.

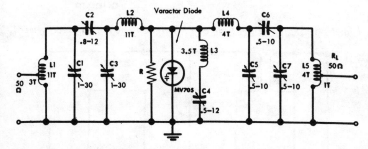

Fig. 10-4. This frequency multiplier circuit is designed to operate as a doubler or quadrupler.

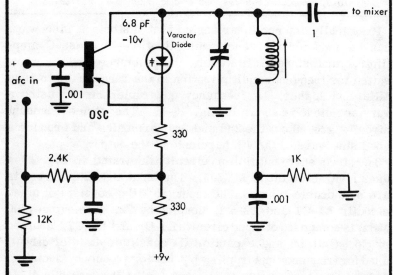

Fig. 10-5. This recently developed AFC circuit provides control over a wide range and is more sensitive to drift.

Multiplying Ratios for Fewest Diodes in Doubler, Tripler, and Quadrupler Cascades—

Output in Gc. (0.1 to 0.2 Gc Input)	Mult. Ratio	Fewest Mult. Factors
0.2 to 0.4	2	2
0.3 to 0.6	3	3
0.4 to 0.8	4	4
0.6 to 1.2	6	2 × 3
0.8 to 1.6	8	2 × 4
0.9 to 1.8	9	3 × 3
1.2 to 2.4	12	3 × 4
1.6 to 3.2	16	4 × 4
1.8 to 3.6	18	2 × 3 × 3
2.4 to 4.8	24	2 × 3 × 4
2.7 to 5.4	27	3 × 3 × 3
3.2 to 6.4	32	2 × 4 × 4
3.6 to 7.2	36	3 × 3 × 4
4.8 to 9.6	48	3 × 4 × 4
6.4 to 12.8	64	4 × 4 × 4

circuit is capable of delivering 22 watts of power output when used in conjunction with a 50-MHz transistor transmitter.

FREQUENCY MULTIPLICATION IN THE GIGACYCLE RANGE

Varactors are performing an important function in transistorized microwave generators since they permit frequency multiplication without excessive power loss. Consequently, they are being considered as a replacement for klystron tubes in at least some applications. The minimum number of varactor diodes required in cascading multiplier circuits to raise 0.2 gigacycles to 12.8 gigacycles is given in the accompanying Table.

AUTOMATIC FREQUENCY CONTROL

Another area where the varactor shows great promise is in automatic frequency control (AFC) of local oscillators. Most AFC systems presently in use perform satisfactorily, provided the frequency remains within 3% of its normal value. Transistor receivers generally use varactor-type AFC systems with moderate success but many are limited to a very narrow control band and low sensitivity. However, a recently developed varactor system, one that rates above all other types, is shown in Fig. 10-5.

Notice that a 1000-ohm resistor in the collector supply permits the oscillator to function as a DC amplifier between the AFC input and the varactor diode. The varactor also func-

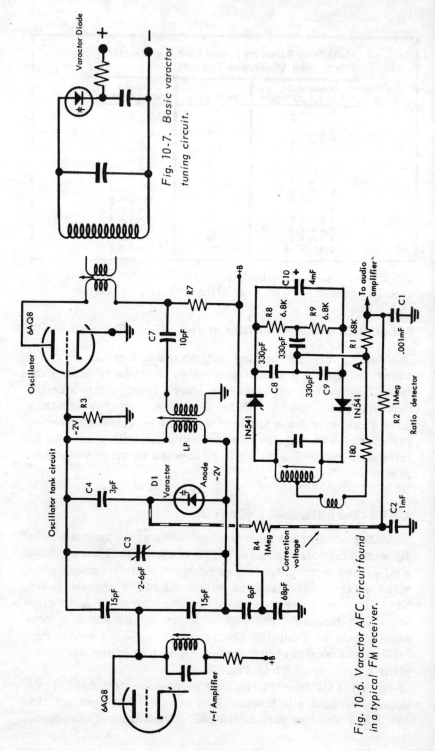

Fig. 10-7. Basic varactor tuning circuit.

Fig. 10-6. Varactor AFC circuit found in a typical FM receiver.

tions as a feedback capacitor, the value of which increases as its Q value decreases; therefore, as the load on the oscillator increases, so does the feedback capacity. With this configuration, a sensitivity of 5.8 MHz is experienced with a tuning range of over 11 MHz in a 40-MHz oscillator. This is ideal for multiband receivers and sweep generators. The oscillator and ratio detector sections of a typical FM receiver, using a varactor for oscillator frequency control, is illustrated in Fig. 10-6.

The two prime requirements for effective automatic frequency control are a reactive component sensitive to a DC voltage and a frequency discriminator to provide a correction voltage. In an FM receiver the filtered output of the ratio detector serves as a frequency discriminator and provides a DC correction voltage. Notice in Fig. 10-6 that the audio take-off from the ratio detector is Point A. This is followed by a de-emphasis network R1-C1 and fed to the first audio stage. The audio line then proceeds through the R2-C2 filter which removes the audio and develops a DC voltage that is proportional to any oscillator drift that may occur.

The varactor anode is connected to the control grid of the oscillator through coil LP and picks up a negative 2 volts from the (self-bias) oscillator grid. The oscillator tank circuit consists of C3 and LP, shunted by C4 and varactor D1. The correction voltage is applied as a control bias to the cathode of the varactor via R4. Now, when the receiver is tuned properly and the oscillator frequency tracks correctly, the correction voltage at the cathode of D1 is zero; hence, the total bias across D1 is 2 volts negative at the anode. Should the oscillator drift to a higher than normal frequency, the cathode of D1 is biased negatively. This decreases the total reverse bias and narrows the depletion width, which in turn increases the tank circuit capacitance and lowers the frequency. When the oscillator drifts to a lower frequency, the correction voltage is positive at the cathode which increases the reverse bias and depletion width. This decreases the tank circuit capacitance and increases the frequency. The varactor diode is about a 1/4" long and a 1/16" in diameter. The small size is due to the dielectric constant of silicon which is 12 as compared to 1 for air. Also, the dielectric widths of the depletion zone are very narrow.

The varactor is used also as a tuning circuit in radio and TV

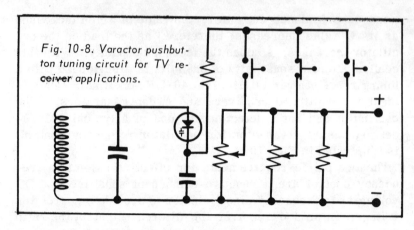

Fig. 10-8. Varactor pushbutton tuning circuit for TV receiver applications.

receiver circuits, a technique quite adaptable to remote location of the tuning control since tuning is accomplished simply by varying the voltage input to the circuit. Fig. 10-7 is a basic varactor tuning circuit and Fig. 10-8 is a pushbutton tuning system suitable for a TV tuner.

Chapter 11

High-Frequency Oscillators and Amplifiers

Semiconductors are widely used and are finding increasing application in various high-frequency oscillator and amplifier circuits. Several are considered here.

30-MHz OSCILLATOR

The base bias current is obtained from voltage-divider R1-R2 which operates in conjunction with RF choke coil L1 and filter capacitor C1. See Fig. 11-1. Resistance R3 connected in the emitter lead provides DC stability and is bypassed by capacitor C3 to prevent degeneration. All the collector current flows through R3; therefore, any excessive shift in this current due to a temperature is opposed by the emitter bias developed across R3. This is DC negative feedback and, since the emitter is at AC ground potential (due to C3), the signal is not affected.

The collector load consists of a tuned tank circuit (C4, C5, T1). T1 functions as an autotransformer in which the top four turns represent the primary, and the bottom seven turns represent the secondary. Notice that the collector bias voltage is applied through a filter circuit (L2, C7, C8) to the 4-turn tap on T1. Positive feedback for regeneration, coupled from the output of the secondary, is applied to the base through capacitor C2. Both feedback coupling and output capacitors C2 and C6 are adjusted for maximum response. The tank circuit, of course, determines the oscillator frequency.

30-MHz AMPLIFIER

A 30-MHz amplifier using a 2N2188 transistor is illustrated

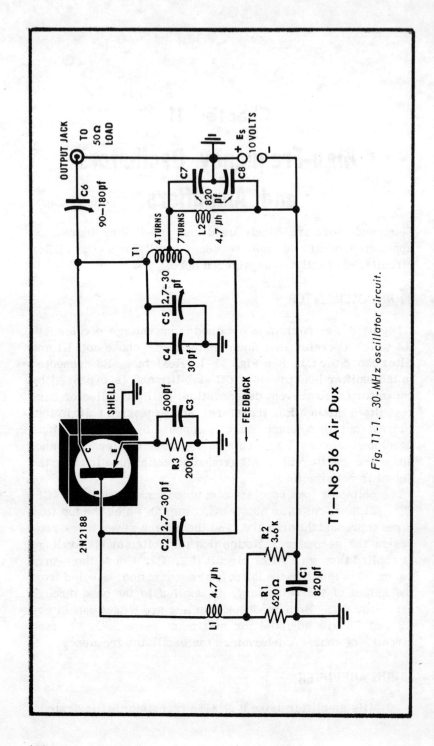

Fig. 11-1. 30-MHz oscillator circuit.

in Fig. 11-2. This circuit is a transistorized version of the tuned grid-tuned plate vacuum tube amplifier. The input tank circuit is tapped 1 1/4 turns above ground to match the output of the oscillator shown in Fig. 11-1. A second tap, a little higher, is used to match the relatively larger base input impedance and is connected to the base through coupling capacitor C1 which serves also as a blocking capacitor to prevent grounding the DC base bias. DC base bias is applied through a voltage divider consisting of R1 and R2. Resistor R3, connected in the emitter lead, is used to provide DC stability and is bypassed to prevent degeneration.

The collector output drives a 30-MHz tank circuit (C3 and the primary of T2). Notice that capacitor C4, which has a low impedance at 30 MHz, effectively places the primary of T2 in parallel with tuning capacitor C3. The blocking action of C4 is necessary to isolate the negative DC supply line from ground. Base tank coil T1 is 7 turns of No. 16 A.W.G. with an outside coil diameter of 0.6 inches and an overall length of 0.8 inches. Tap N1 is located at 5.25 turns; N2 at 0.5 turns; and N3 at 1.25 turns. Output transformer T2 is a Cambridge Thermionic coil PLS 62C4L/20063D, or equivalent. Winding N4 is 12 turns of No. 30 A.W.G. with an inductance of 1.2 μh. N5 is 3 turns of No. 30 A.W.G. with an inductance of 0.18 μh and a coupling coefficient of 0.58.

INTERMEDIATE FREQUENCY AMPLIFIER

The intermediate frequency (IF) amplifier is an integral part of a superheterodyne receiver; located between the "first" and "second" detectors, it contributes practically all the amplification of the received signal prior to demodulation.

The superheterodyne receiver was developed in 1916 and, because of its high sensitivity, was first used in Marine compass stations. The coast station operator was able to determine the approximate position of a ship by rotating a loop antenna and tuning in the transmitter signal. Later three compass stations were set up along the coast so that the position of the ship (latitude and longitude) could be pinpointed by triangulation. Still later, ships carried their own compass stations and used land beacons to determine their positions. Second to the advent of Marine wireless communications, the superheterodyne receiver was another blessing to navigation.

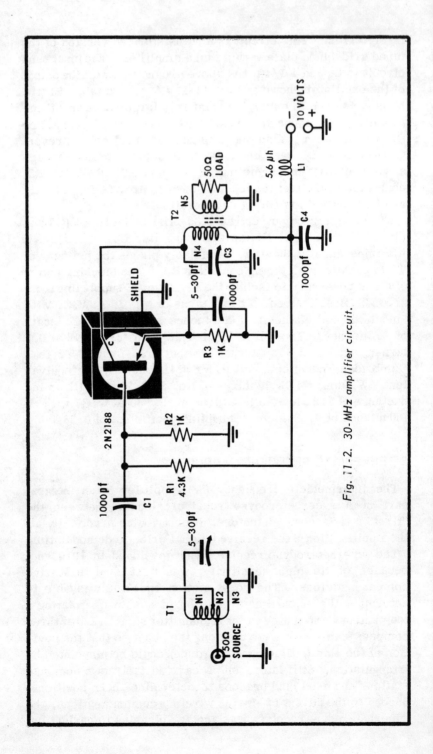

Fig. 11-2. 30-MHz amplifier circuit.

The third boon to navigation was radar which also uses the superheterodyne principle of reception. Since 1916 the super-heterodyne system has not changed but its uses have multiplied with improved circuitry. The superheterodyne principle has been recently adapted to the field of supersonics where it converts these frequencies to audio frequencies. This system is used extensively in the study of wildlife and insect communications in the field of bionics.

It is significant that this almost universally used principle of reception was the brain child of an American engineer, the late Major Armstrong, who also developed the FM system as we know it today, a system which also uses the superheterodyne principle of reception. The popularity of superheterodyne stems from the fact that all radio signals, regardless of frequency, can be converted to a fixed lower frequency by super-imposing another (hetero) force (dyne), usually a local oscillator. (Transistorized variation of the IF amplifier circuits follow in this Chapter.)

TRANSISTORIZED UHF TV TUNER

A highly efficient, low-noise, UHF tuner developed by Texas Instruments is illustrated in Fig. 11-3. The circuit uses three 2N2415 transistors that have a maximum frequency rating of 3 gigacycles. In this tuner design we pass from the coil and loop inductance to a "tuned line" that operates exactly like a resonant transmission line. The line is in the form of a flat strip to provide adequate surface area to lower the RF impedance due to "skin effect." The strip-type inductance with a tuning capacitor functions as a tank circuit and may be tapped at any impedance to provide proper matching. The taps are calculated at 890 MHz to provide maximum power gain and proper "Q" for the required bandpass.

The DC bias network furnishes the operating potentials for the RF amplifier emitter and collector. The base is AC-grounded and the collector is grounded through a portion of the tuned line which constitutes the collector load. The stage has a relatively high voltage gain with excellent stability.

The oscillator is also a common-base stage with a bias network identical to the RF stage. Regenerative capacitive feedback between emitter and collector is obtained by a short length of wire located in close proximity to the collector tuned line.

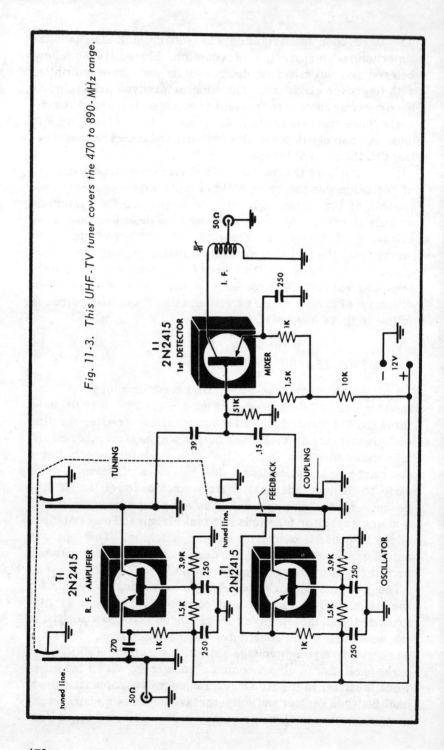

Fig. 11-3. This UHF-TV tuner covers the 470 to 890-MHz range.

The oscillator output is inductively coupled by a short piece of wire which functions as a secondary and this, in turn, is capacitively coupled to the mixer input to prevent grounding the DC bias circuit. The output delivers an RF injection of .4 volt at 470 MHz and .07 volt at 980 MHz.

The mixer is a common-emitter stage and was selected because it provides a noise reduction advantage; however, this does not rule out the common-base stage in subsequent designs. The diode type mixer was abandoned here because of the accompanying loss of overall conversion gain. The use of a low-emitter current of about .1 ma holds the noise figure

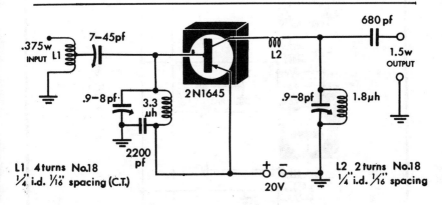

Fig. 11-4. 160-MHz amplifier circuit.

down. The output is tuned to 45 MHz with a 10% bandpass. Total power consumption is 18 ma at 12 volts.

160-MHz AMPLIFIER

A 160-MHz amplifier circuit using a Western Electric 2N1645 transistor is illustrated in Fig. 11-4. This circuit is a simple tuned-base, tuned-collector circuit with a power gain of approximately 4 to 1.

The 2N1645 is a diffused base germanium mesa transistor used for UHF operation and high-speed switching. For example, under constant drive conditions the "on" and "off" elapsed times are 5 and 15 nanoseconds, respectively. As a frequency doubler it can operate up to 250 MHz and provide an output at this frequency of .5 watt. Frequency doubling is usu-

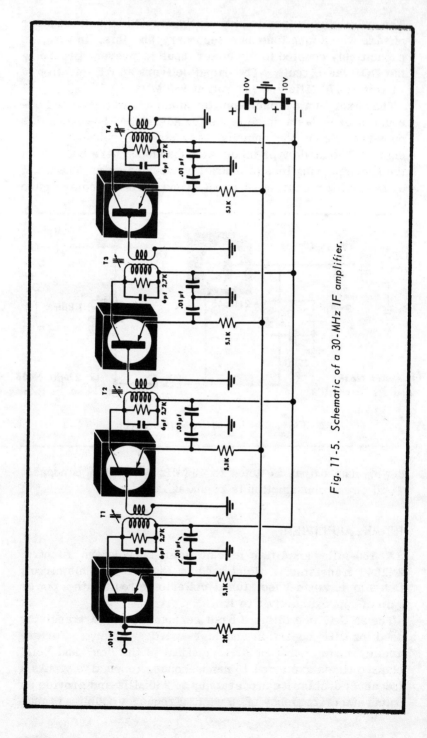

Fig. 11-5. Schematic of a 30-MHz IF amplifier.

ally accomplished by tuning the collector tank circuit to twice
the input frequency.

30-MHz IF AMPLIFIER

Fig. 11-5 is a 30-MHz, four-stage IF amplifier circuit using
four transistors and four single-tuned coupling transformers.
Separate voltage supplies are used for emitter and collector
feeds to provide good bias stability.   Each tank circuit has a
resistive load (27K) to broaden the bandwidth, which is approximately 3.5 MHz.   Although this limits the overall gain, it
eliminates the need for neutralization.

To reduce drifts due to temperature variations, each tank
circuit has a capacitive load (6 pfd), hence the tuning does
not rely solely on the inherent capacity of the transistor, which
could otherwise be troublesome. The amplifier is capable of
operating over a temperature range of $-40^{\circ}C$ to $+65^{\circ}C$ with a
gain of better than 80 db. The circuit is an example of solid-
state IF amplifier simplicity and stability.

10.7-MHz IF AMPLIFIER

Fig. 11-6 is a 10.7 MHz IF amplifier designed for an FM
radio receiver. The circuit consists of four transformer-
coupled common-base amplifier stages. The common-base
connection is noted for its excellent stability and high voltage gain.   The stage has a low-impedance input and high-
impedance output, and matching is accomplished by an impedance stepdown transformer. Notice the emitter-input
impedance is approximately 10 times lower than the common-emitter connection.

The input signal is applied from the secondary to the emitter
via a small coupling capacitor. This parallel feed arrangement reduces the input capacity and provides a convenient
method for applying DC bias to the emitter. Each primary is
tuned to the center frequency by its respective trimmer capacitor. This capacitive load makes the tuned circuit somewhat independent of the inherent and stray capacitance of the
transistor.

Each emitter circuit is bypassed to ground by capacitor CF
and an RF choke is used in conjunction with this capacitor to
provide a decoupling circuit between each emitter and the

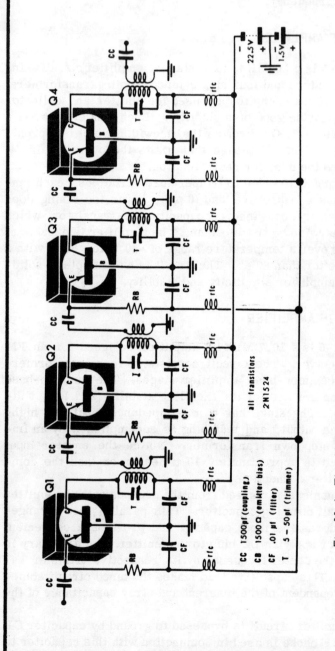

Fig. 11-6. This 10.7-MHz amplifier circuit was designed for FM receiver application.

CC 1500pf (coupling)
CB 1500 Ω (emitter bias)
CF .01 µf (filter)
T   5 – 50 pf (trimmer)

all transistors
2N1524

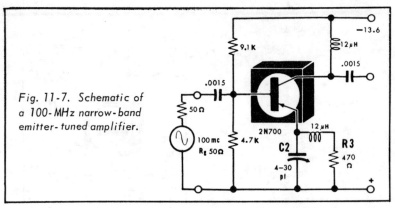

Fig. 11-7. Schematic of a 100-MHz narrow-band emitter-tuned amplifier.

common bias supply line. It is evident that the signal voltage on each emitter is higher than that on the preceding emitter and, since all emitters return to the same bias supply line, there would be a tendency of interaction from the higher signal level to a lower level if decoupling circuits were not used. A similar filter and decoupling arrangement also is provided to each collector circuit. There is NO phase inversion throughout the four stages; therefore, the signal on each emitter is of the same phase but with increasingly higher voltage amplitudes.

RC-COUPLED HIGH-FREQUENCY AMPLIFIER

Use of the inherent reactive properties associated with transistor electrodes and terminal leads has simplified circuit design in high-frequency amplifiers. Although the employment of these inherent reactances in vacuum tubes has become common practice, their use in transistor amplifiers is comparatively new. The recently discovered principle of emitter tuning has made possible improved techniques in the design of high-frequency transistor receivers and wideband amplifiers in the VHF and UHF ranges. The principle of emitter tuning has resulted in a greatly increased transconductance by eliminating degeneration due to inherent emitter inductance. Thus, a small collector load resistance is sufficient, and the improved circuit stability enhances the possibilities of microminiaturization.

An RC-coupled single-stage amplifier incorporating emitter tuning is illustrated in Fig. 11-7. Notice that a small variable capacitor (C2) appears in the emitter lead, forming a series resonant circuit with the inherent emitter inductance.

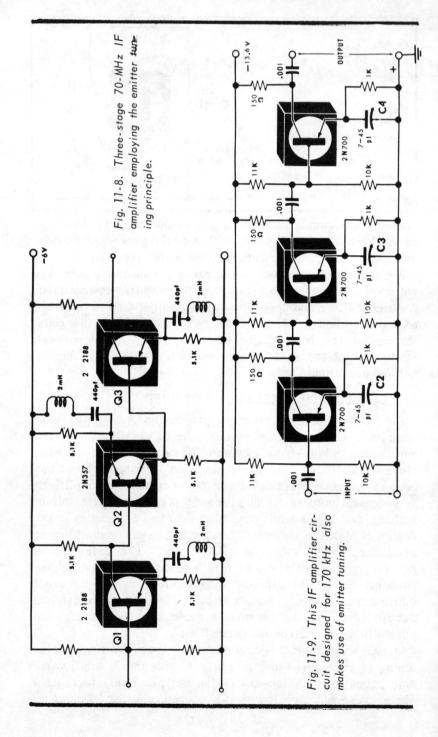

Fig. 11-8. Three-stage 70-MHz IF amplifier employing the emitter tuning principle.

Fig. 11-9. This IF amplifier circuit designed for 170 kHz also makes use of emitter tuning.

The center frequency of this circuit is 100 MHz. To prevent the lowering of the circuit Q, a choke is connected in series with the emitter bias resistance (R3). Eliminating the emitter-inductive degenerative effect by using it as an inductance in the series resonant circuit does raise the input resistance and reduces the output resistance of the transistor. This is an advantage since it reduces the mismatch ratio between cascaded amplifiers.

A 3-stage RC-coupled high-frequency amplifier is illustrated in Fig. 11-8. This amplifier is designed for wideband operation—approximately 35 MHz with a center frequency of 70 MHz. Omitting the RF choke in each stage permits this wider bandwidth. However, the emitter bias resistance of each stage has been increased to 1000 ohms to reduce the shunting effect that would exist without the choke coil.

The 2N700 germanium transistors were replaced with 2N834 silicon types for comparison. This resulted in less gain and an increase in the mismatch ratio due to the lower input resistance of the 2N834. An equivalent 3-stage RF amplifier using tuned LC stages is capable of a gain of 30 to 45 db using 2N700 transistors. However, the increased number of components and complex wiring gives way to the simpler RC-coupled amplifier in high-frequency applications.

A 170-kHz emitter-tuned IF amplifier is illustrated in Fig. 11-9. This interesting departure from the conventional tuned amplifier features a narrow bandwidth and is capable of operating with any general purpose transistor. The advantages of this design are:

1. Simple circuitry
2. Excellent stability
3. Easy to align

Each stage is DC-coupled, thereby eliminating coupling networks. A NPN transistor is used as an intermediate stage to provide correct base-bias polarity. Direct coupling between the collector and base develops a mismatch which eliminates the need for neutralizing circuits as well as providing excellent stability. Although these are desirable features they are obtained by sacrificing gain.

The emitter resistors shunting the series-tuned circuits are used primarily as a DC path to each emitter; however,

they do broaden the bandwidth and should be made as large as
the supply voltage will permit. Bandwidth is also dependent
upon the LC ratio of the series-tuned circuits, and it decreases
as the LC ratio increases. The "Q" of the coil is another
important factor since it effects the gain more than the bandwidth; therefore, considering these inter-related characteristics, a coil with a low value of "L" and a high "Q" is rec-
ommended.

Chapter 12

Field-Effect Transistors (FET)

The field-effect transistor (FET) is the only solid-state device approaching the characteristics of the vacuum tube. The construction, terminology, and principles involved are different from the conventional transistor. Terminology includes such terms as "source," "drain," and "gate," which are the vacuum tube equivalents for "cathode," "plate," and "grid," respectively.

Simply stated, the FET amplifier has a high input impedance and a relatively low output impedance. An amplifier using a field-effect transistor is illustrated in Fig. 12-1. Notice the similarity to a vacuum tube circuit. Despite the large load resistance, the amplifier has only a 20- to-1 voltage gain. This low gain factor is due to the low mutual conductance (Gm) of the FET. For example, the Type C610, manufactured by "Crystalonics" has a Gm of 100 micromhos; however, higher values of Gm are available.

The high input impedance is due to the reverse bias applied to the "gate-source" where the input current consists of minority carriers and not majority carriers. Therefore, a small amount of power in the input circuit develops a relatively large amount of power in the output circuit. The high input impedance characteristic of the FET may be used to advantage and provide more gain when used in conjunction with a conventional transistor as shown in Fig. 12-2. Furthermore, composite units using such combinations have produced mutual conductances ranging higher than 18,000 micromhos. It has been stated that the field-effect transistor provides a conventional transistor with a high-impedance "front end." See Fig. 12-2.

A VOLTAGE-CONTROLLED NONLINEAR RESISTANCE

A field-effect transistor may be used to vary the frequency of a pulse generator without varying the pulse width. The field-effect transistor, when used in a pulse generator as a control component, provides greater reliability to a conventional multivibrator. The circuit illustrated in Fig. 12-3 shows a basic emitter-coupled multivibrator, and in Fig. 12-4 is the same circuit using a field-effect transistor for frequency control. The relatively high input resistance of the field-effect transistor permits frequency control by a voltage signal obtained from a high internal impedance source. A

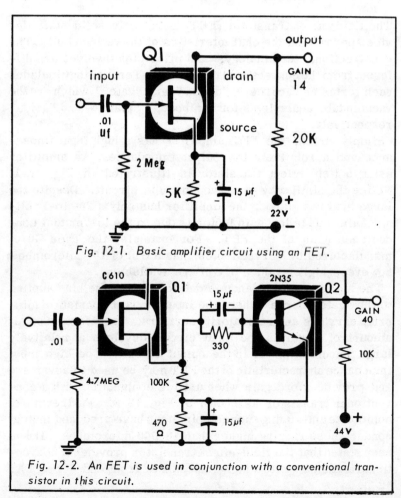

Fig. 12-1. Basic amplifier circuit using an FET.

Fig. 12-2. An FET is used in conjunction with a conventional transistor in this circuit.

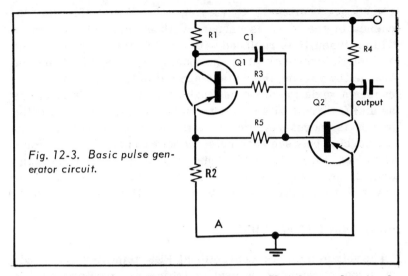

Fig. 12-3. Basic pulse generator circuit.

study of the basic circuit in Fig. 12-3 will aid in understand-
ing the operation of the circuit in Fig. 12-4.

When power is applied capacitor C1 begins to charge through
R1 and the base-emitter junction of Q2, causing Q2 to conduct and eventually saturate. At this point the collector of
Q2 is practically at ground potential, which in turn causes
Q1 to cut off. This condition prevails while C1 is charging,
thus holding Q2 at cut off for a period determined by the time
constant of R1-C1. As capacitor C1 approaches full charge,
the current through the base-emitter junction of Q2 falls below a value required to keep Q2 conducting. At this moment
Q2 is cut off and the negative voltage at its collector increases
rapidly and drives the base-emitter junction of Q1 through R3,
which in turn causes Q1 to conduct and saturate. The collector voltage of Q1 decreases and C1 discharges through R5
and the conducting transistor (emitter-collector path). The
discharge current of C1 flows through R5 and reverse biases
the base-emitter junction of Q2. The voltage drop across R5
is greater than the voltage drop across R2; however, as C1
continues to discharge, this voltage ratio is reversed and Q2
starts to conduct, which in turn causes Q1 to cut off. At
this moment, C1 starts to recharge and the regenerative
cycle is repeated.

Disregarding the internal resistance of the conducting transistor, capacitor C1 is part of two separate RC time constants; i.e., R1-C1 while charging, and R5-C1 when dis-

charging. Therefore, by varying either R1 or R5, the frequency of the pulse is changed without disturbing the pulse width. When R5 is replaced with a field-effect transistor and an isolation resistor (R6), the spacing between pulses may be changed by varying the voltage applied to the FET gate. This causes a variation of its internal resistance which functions as an RC circuit with C1. Therefore, as the frequency of the pulses is increased, the pulse width remains unchanged but the spacing between each pulse is decreased. See waveform 2f in Fig. 12-4. The control may be classified as a "voltage-controlled, nonlinear resistor." With high-speed switching transistors, frequencies up to several MHz can be obtained.

HIGH-GAIN FET AMPLIFIER

The voltage gain of a conventional FET amplifier is dependent upon the value of the load resistor—the greater its value the greater the gain. Of course, the value is limited by the available voltage supply. For example, a load resistor of 20K and an operating potential of 15 volts will provide a voltage gain of approximately 20 to 1. See Fig. 5-1. Let us consider using a second FET stage Q2 connected in place of

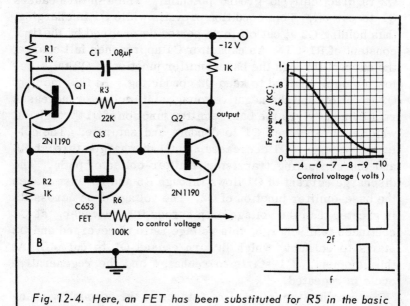

Fig. 12-4. Here, an FET has been substituted for R5 in the basic circuit (Fig. 12-3).

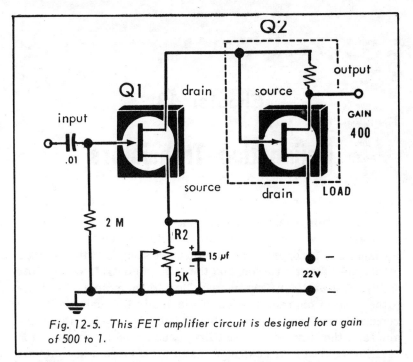

Fig. 12-5. This FET amplifier circuit is designed for a gain of 500 to 1.

the 20K load resistor, with its gate electrode connected to the DRAIN TERMINAL OF Q1. See Fig. 12-5.

When a positive-going signal is applied to the gate of Q1 it causes the DRAIN current to increase and the DRAIN voltage to become less positive (negative-going). This negative excursion of the signal is applied to the gate electrode of Q2, reducing its DRAIN current. This action is tantamount to an increase of internal resistance, since the Q2 input is the output load resistance of Q1. The increase in resistance is very substantial due to the amplification properties of Q2. However, when the signal applied to Q1 is negative-going it has the reverse effect on Q2 which is a substantial reduction of the load resistance. The applied 15 volts is equally distributed between Q1 and Q2 by adjusting the potentiometer R2. Hence the operating voltage (with no signal) on each stage is approximately half the supply voltage. When a 5-millivolt signal is applied to the gate of Q1, the amplified signal at the output (Q2) will be approximately 2.5 volts, which is a gain of 500 to 1.

Chapter 13

Unijunction Transistors

The unijunction transistor, developed by the General Electric Company, is a 3-terminal semiconductor device with a single PN junction. Its operating principles are similar to the gaseous discharge tube (thyratron) and the silicon controlled rectifier (SCR). The unit was first named "double base diode," but due to the control characteristics of the single PN junction it was renamed "unijunction transistor." See Fig. 13-1. As an example of the device's versatility, when used as a relaxation oscillator, the circuit can be adjusted to produce a near linear sawtooth waveform. An oscillator circuit with a resistor connected in series with each base lead produces three different waveforms simultaneously. See Fig. 13-2.

UNIJUNCTION APPLICATIONS

Application of the unijunction transistor is not limited to pulse work but can be used to advantage in many other applications, thanks to the following attributes:

• Operable over a wide temperature range
• High sensitivity and excellent stability
• Frequency, amplitude, and pulse width are easy to control
• Negative resistance characteristic
• Ideally suited for control circuits
• Capable of free-running, bistable, monostable operation
• Low cost: basis for integrated circuitry

Operation

Let us assume that the applied potential between Base 1 and

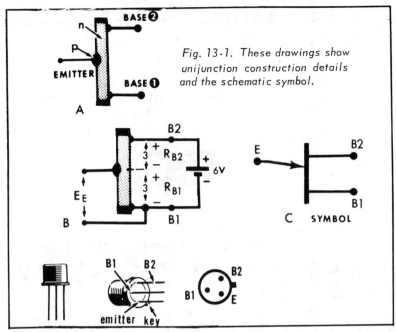

Fig. 13-1. These drawings show unijunction construction details and the schematic symbol.

Base 2 is 6 volts, and the P-type emitter is located mid-way between the two base terminals, as illustrated in Fig. 13-1B. The voltage drop across the bar produces a uniform voltage gradient through its internal resistance R_{BB} which has a resistance measurement between 5K and 10K; therefore, midway along the silicon bar (in the vicinity of the emitter,) there will be a potential of 3 volts with respect to Base 1. When the emitter lead is connected to Base 1 the PN junction has a reverse bias; i. e., the N-type bar is +3 volts with respect to Base 1, making the P-type emitter -3 volts with respect to the N-type bar. The reverse bias across the PN junction results in emitter current cut-off and the transistor will remain in this condition until fired.

To forward bias the junction requires an external potential of at least +3 volts applied to the emitter with respect to Base 1, which is the threshold of conduction, and a slight increase above this value causes a relatively large current to flow between the emitter and Base 1, which is tantamount to a reduction in resistance between these two points. When this occurs, the transistor is said to be "fired." It is correct to assume that the positive bias applied to the emitter injects positive carriers from the P-type area into the N-type bar, causing the transistor to fire. In practice the emitter is located off center toward Base 1, thus requiring a lower firing point.

Circuitry

A sawtooth oscillator using a unijunction transistor is shown in Fig. 13-3. When the power switch SW1 is closed, the PN

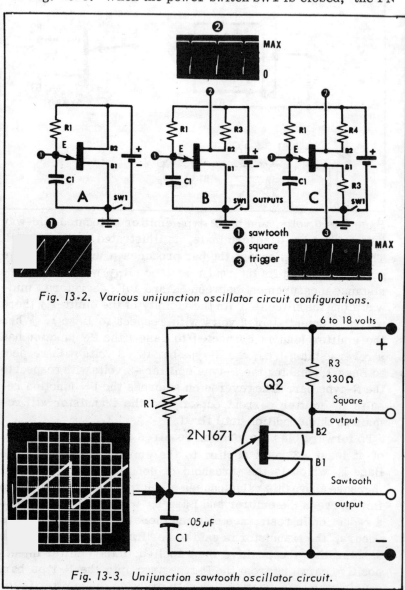

Fig. 13-2. Various unijunction oscillator circuit configurations.

Fig. 13-3. Unijunction sawtooth oscillator circuit.

junction—at this moment—is reversed biased and capacitor C1 starts to charge. Notice that C1 is connected between Base 1 and the emitter, and the charging current develops a forward bias (integrator voltage) across the capacitor. When this potential approaches the value of the existing reverse bias, the threshold voltage is reached and the transistor fires. The time taken for the capacitor voltage to reach the firing point depends upon the RC time constant of C1-R1.

When the junction conducts, C1 discharges rapidly through the low-resistance path between Base 1 and the emitter. The voltage across C1 falls rapidly and the transistor is again reverse-biased and cut off. This completes one cycle of capacitor charge and discharge, which is repeated at a frequency determined by the time constant of R1-C1. The sawtooth waveform that represents the charge and discharge curve of C1 appears at the emitter terminal, and since this waveform is recurrent, the circuit may be classified as "free-running."

THREE OUTPUTS: SAWTOOTH, SQUARE WAVE AND TRIGGER

Fig. 13-2B shows a modified circuit that provides both sawtooth and square wave. To accomplish this, load resistor R3 is connected in the Base 2 lead. The circuit operates in the same manner as that in Fig. 13-3; however, the addition of R3 develops a square wave of voltage at the Base 2 terminal. The addition of another resistance in series with the Base 1 lead produces a positive trigger pulse. See Fig. 13-2C. The three outputs appear simultaneously and may be used independently for various applications.

FREE-RUNNING MULTIVIBRATOR

The multivibrator circuit illustrated in Fig. 13-4 incorporates a diode (D1) and a shunt resistor (R2) connected across the capacitor. The moment power is applied the transistor is reverse-biased and C1 starts to charge through the diode (D1) and R1. The difference between this circuit and those previously discussed is the maximum voltage to which C1 can charge, a parameter determined by the voltage ratio R1/R2. Notice that R2 is 8K and R1 is 10K, hence the drop across R2 is the maximum voltage to which C1 can charge. However, the voltage across C1 must be sufficient to fire the transistor. When firing occurs, the diode is cut off and the transistor remains fired for a period determined by the time constant of R2-C1.

The operation of the diode may be explained as follows: At the moment of firing, the emitter voltage drops close to ground potential, developing a reverse bias across the diode; i.e., the diode cathode is more positive than the anode, and an open circuit (isolation) exists between points "a" and "b" (see Fig. 13-4). During the diode cut-off period the transistor remains fired and a current (electrons) flows from the negative line via Base 1, the PN junction, and R1 to the positive line. In the interim, C1 is discharging through R2, and the voltage drop across this resistance gradually decreases. Eventually, the cathode of D1 becomes negative with respect to its anode and the diode conducts. When this occurs, the transistor is again reverse-biased and C1 begins to recharge and the cycle is repeated. The "off" period of

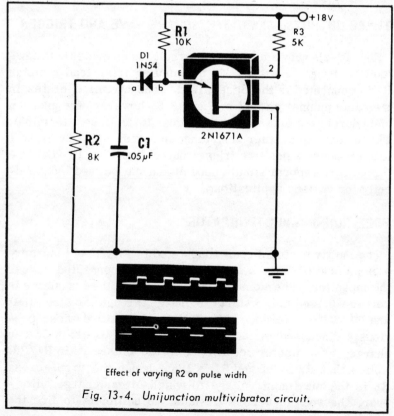

Effect of varying R2 on pulse width

Fig. 13-4. Unijunction multivibrator circuit.

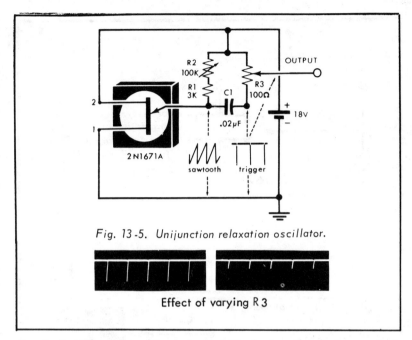

Fig. 13-5. Unijunction relaxation oscillator.

Effect of varying R3

the diode determines the pulse width during the firing period, the duration of which is dependent upon the constant of R2-C1.

UNIJUNCTION TRIGGER GENERATOR

A unijunction transistor oscillator differing somewhat from the relaxation oscillator is illustrated in Fig. 13-5. The difference is in the charge sequence of capacitor C1. Notice the charge path is from the negative line via Base 1, emitter, and R3, to the positive line. When power is applied, the voltage distribution across the charge path is such that it develops a forward bias across the PN junction. This causes an emitter current to flow through the charge path until C1 is charged. When this point is reached, the emitter current is almost zero and the capacitor discharges through R1, R2, and R3.

The time constant of the charge path is relatively short and C1 charges rapidly, developing a negative trigger pulse at the junction of C1 and R3 with respect to ground. The time constant of the discharge circuit is longer, permitting a sawtooth wave to appear at the junction of C1 and R1; during this time the emitter is cut off. Comparing this oscillator to the relaxation type, it is interesting to notice that the transistor

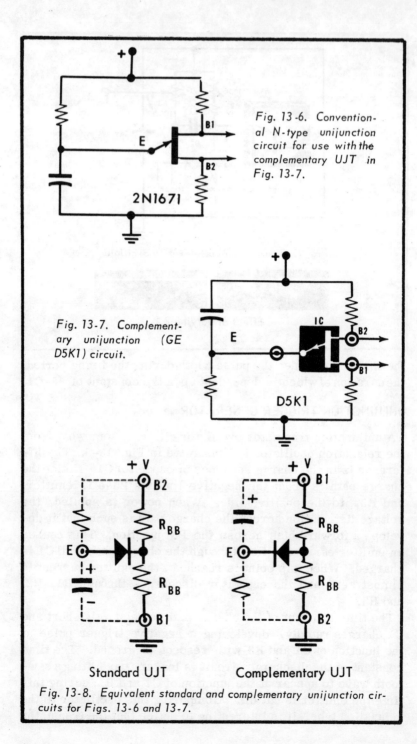

Fig. 13-6. Conventional N-type unijunction circuit for use with the complementary UJT in Fig. 13-7.

Fig. 13-7. Complementary unijunction (GE D5K1) circuit.

Standard UJT

Complementary UJT

Fig. 13-8. Equivalent standard and complementary unijunction circuits for Figs. 13-6 and 13-7.

is fired first and its emitter current charges C1. This action is the reverse of the relaxation oscillator previously discussed.

COMPLEMENTARY UNIJUNCTION TRANSISTOR (UJT) GE DSK1

Integrated-circuit fabrication techniques have been used to develop and improve unijunction transistors. Although a discrete unit, it has a higher cut-off frequency and remarkable stability. Like the standard UJT, the complementary UJT is a relaxation oscillator with a few minor circuit changes. For example, the standard UJT is an "N" type and the complementary UJT is a "P" type material between B1 and B2 and uses a reverse polarity to that of the standard UJT, thereby requiring a reversal of its two base resistors. Also, the switching action of the complementary UJT is different. See Figs. 13-6 and 13-7 for comparison and the equivalent circuits in Fig. 13-8.

The conventional UJT has an alloyed "PN" junction that conducts the total transistor modulation current when triggered. The complementary UJT has a PN-PN switch action similar to the silicon controlled switch and draws a relatively small trigger current. All these improvements, plus monolothic construction, makes the complementary UJT a desired unit for use in television sweep circuits and other systems requiring accurate timing pulses. The complementary unit is fabricated to consist of a PN-PN switch and two diffused resistors between B1 and B2. The General Electric complementary UJT (designated D5K1) is shown in Fig. 13-7. The two types function as an RC control oscillator; however, the complementary functions better.

Chapter 14

Zener and Special Purpose Diodes

When a specified reverse-bias voltage (zener voltage) is applied to a zener diode, it causes the diode to breakdown and establish a fixed voltage, thus making it useful for regulatory purposes. For example, a General Electric zener diode Z4XL6.2B will breakdown and establish a 6.2-volt reference level.

The breakdown may be defined as a spontaneous disruption of covalent bonds between the atoms in the vicinity of the PN junction. This condition is due to the reverse-bias voltage that produces an electric field sufficiently strong to excite and release electrons in the depletion zone. When this occurs an avalanche of current flows through the diode, limited only by the resistance of the external circuit. It is interesting to note that the voltage across the diode after breakdown remains constant at the specified zener voltage.

Increasing the reverse-bias potential beyond the zener voltage causes the diode current to increase while the voltage drop across the diode remains relatively steady. This nonconformity with Ohm's law is due to the decreasing internal resistivity of the diode as the current through it increases, thus holding a steady IR drop across the diode. It is this characteristic of the zener diode that makes it ideal for regulatory devices or voltage reference levels.

A 20-volt, 1-watt zener diode test setup is shown in Fig. 14-1A. Notice that the line voltage (117v) is applied through a 6-watt lamp to the test circuit for protection against an accidental short circuit. The test circuit provides a synchronized 60-Hz sweep excursion of the horizontal and vertical sweep of the oscilloscope. The AC voltage also is applied to

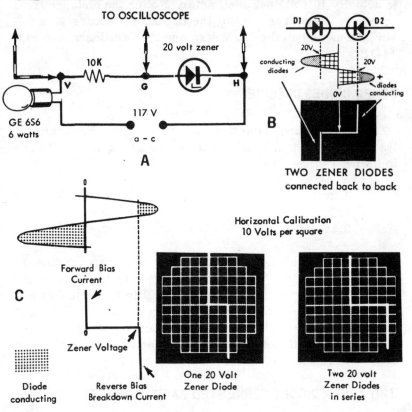

Fig. 14-1. Zener diode test circuit with test curves.

the test circuit through a 10K resistor which provides a forward and reverse bias swing to the diode under test. See Fig. 14-1C. The vertical input to the oscilloscope is connected across the 10K resistor, hence the reverse-bias voltage across the diode will be represented by a flat trace which indicates zero current through the series 10K resistor.

It must be remembered that the AC horizontal sweep voltage is zero at the center of the horizontal trace. Therefore, the horizontal sweep represents the reverse-bias potential or zener voltage since the flat trace indicates the diode is not conducting. Therefore, the horizontal sweep represents a reverse-bias, half-cycle swing. Notice the breakdown (zener voltage) occurs at a distance from center in the reverse-bias region, as illustrated in the oscillogram. In the test, the horizontal sweep was calibrated to 10 volts per square, which is actually 10 volts per centimeter. Notice the distance from center is 2 squares, hence the breakdown occurs at a reverse-bias potential of 20 volts. See the Oscillogram in Fig. 14-1B.

ZENER DIODES IN SERIES

Zener diodes are obtainable at most voltages. For example, the G.E. Z4XL family of zener diodes are available in the following ratings:

6.2v	12v	18v
7.5v	14v	20v
9.1v	16v	22v

Since these are all 1-watt diodes, they may be connected in series to provide a stable voltage equal to the sum of the zener rating for each diode; i.e., a 12v in series with a 20v will provide a zener voltage reference of 32 volts. The addition of a 14v diode increases the zener potential across the series circuit to 46 volts. See the series-connected zener diode circuits in Fig. 14-2.

TWO ZENER DIODES CONNECTED BACK-TO-BACK

Two 20-volt zener diodes connected back-to-back, as shown in Fig. 14-1B, provide a 40-volt peak-to-peak reference level. For example, during the positive halfwave excursion of the input cycle, D2 is forward biased and D1 is reverse biased—neither conducting. However, when the positive halfwave reaches 20 volts, D1 conducts through D2 and maintains a 20-volt drop across both diodes. This continues until the halfwave of the input cycle drops to 20 volts.

During the negative halfwave excursion, D1 is forward biased and D2 is reverse biased—neither conducting. However, when the negative halfwave reaches 20 volts, D2 conducts through D1 and maintains a 20-volt drop across both diodes. The two zener voltages provide a 40-volt sweep as illustrated in the oscillogram. The flat portions of the waveform represent both diodes nonconducting, or zener level. This is a function similar to a thyrector used for clipping AC line transients. For example, any inductive surges above

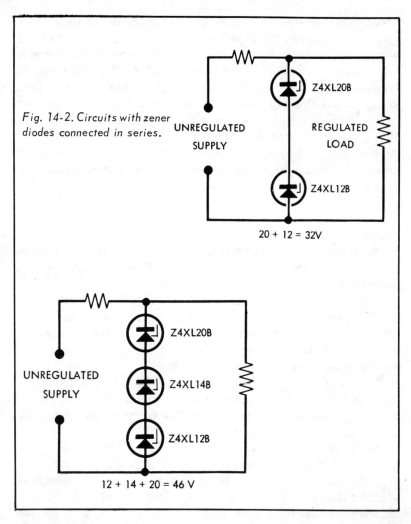

Fig. 14-2. Circuits with zener diodes connected in series.

UNREGULATED SUPPLY

REGULATED LOAD

Z4XL20B

Z4XL12B

20 + 12 = 32V

Z4XL20B

Z4XL14B

Z4XL12B

UNREGULATED SUPPLY

12 + 14 + 20 = 46 V

the breakdown voltage are absorbed by conduction of this bi-directional unit.

TUNNEL DIODE

The tunnel diode is a single PN junction that operates on all frequencies up to 1000 MHz, with a low-noise figure and low-operating power. The essential difference between the tunnel diode and the conventional diode is the high conductivity of the P and N materials used in its construction. The conductivity is more than 1000 times greater than that of conventional diodes. This is due to a greater concentration of impurities (doping) applied to the crystals while forming. The result is a very narrow junction (depletion layer) between the P and N electrodes. The narrow junction permits the electrons to "tunnel" through the potential barrier at a relatively low-bias voltage (less than 50 millivolts for germanium), hence, the name "tunnel." In conventional diodes the electrons lack the energy to penetrate, or overcome, the potential barrier at such a low potential. Actually, with a narrow junction a relatively small bias can develop a field strong enough to cause valence electrons to escape across the barrier at a velocity close to the speed of light.

A "forward" or "reverse" bias of less than 50 millivolts applied to the diode causes the valence electrons of the semi-conductor atoms near the junction to tunnel in either direction in accordance with the bias polarity. If the forward bias is increased more than 50 millivolts, the free electrons in the N region acquire more energy than the valence electrons in the P region, thus causing the tunnel current to decrease. See Fig. 14-3. A decrease in current, despite an increase in voltage, indicates a negative resistance characteristic. However, a further increase of forward bias, above 300 millivolts (for germanium), energizes the free electrons and holes sufficiently to overcome the potential barrier. When this occurs the tunnel diode is similar to the conventional diode. See the dashed line in the sketch (Fig. 14-3).

Despite the bi-directional characteristic of the tunnel diode, it does exhibit a negative-resistance characteristic in one direction as illustrated. It is this region that determines whether the diode is normal or defective. A test circuit may be devised where the applied bias voltage (millivolts) is plotted against tunnel diode current (milliamperes); however, this requires two meters and is time consuming. Hence, the ultimate is a curve tracer circuit and an oscilloscope which provides an excellent display for observation and quantitative analysis. The author experimented with several methods and found the circuit illustrated in Fig. 14-1C to be inexpensive and adequate for the purpose.

Testing a conventional semiconductor diode with an ohmmeter to determine its forward and reverse resistance ratio is both simple and expedient. However, the ohmmeter test must not be used on the tunnel diode; first, the tunnel diode

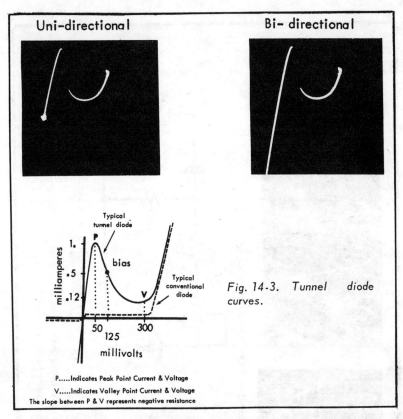

Uni-directional

Bi-directional

Fig. 14-3. Tunnel diode curves.

Typical tunnel diode

P

bias

Typical conventional diode

V

milliamperes

1.
.5
.12

50   300
125

millivolts

P.....Indicates Peak Point Current & Voltage
V.....Indicates Valley Point Current & Voltage
The slope between P & V represents negative resistance

conducts equally in both directions and a resistance test serves no useful purpose; second, damage to the diode can result, especially on a low-resistance (high-current) range.

TUNNEL DIODE OSCILLATOR

The negative resistance of the tunnel diode at a low-forward bias makes it an ideal device for oscillators. An oscillator circuit using a tunnel diode is shown in Fig. 14-4B. Here, the diode functions to replenish the tank circuit losses, thereby sustaining oscillations.

To determine how the diode operates, let us assume that a tuned circuit has no heat-producing resistance and that its oscillatory current produces a magnetic field that returns all its energy to the circuit upon collapse. This would be an ideal condition since one shock pulse would cause the tuned circuit to oscillate continuously—a form of perpetual motion. However, in practice this is not possible since all tuned cir-

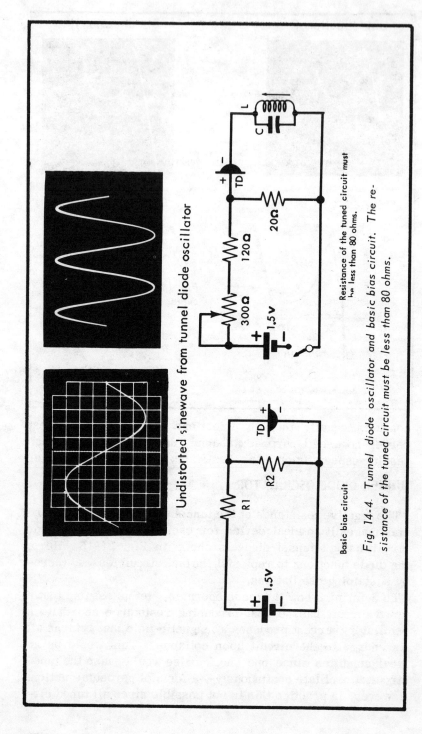

Undistorted sinewave from tunnel diode oscillator

Resistance of the tuned circuit must be less than 80 ohms.

Basic bias circuit

Fig. 14-4. Tunnel diode oscillator and basic bias circuit. The resistance of the tuned circuit must be less than 80 ohms.

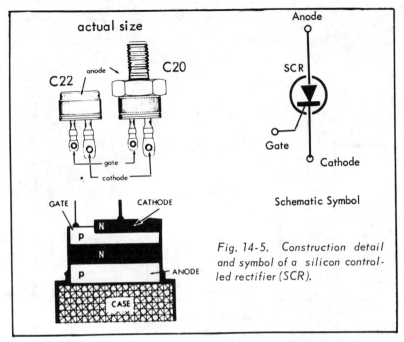

actual size

C22

C20

anode

gate

cathode

GATE   CATHODE

N

P

N

P   ANODE

CASE

Anode

SCR

Gate

Cathode

Schematic Symbol

Fig. 14-5. Construction detail and symbol of a silicon controlled rectifier (SCR).

cuits contain heat-producing resistance (positive resistance), and when shock excited would produce a damped-wave in which each succeeding cycle diminishes in amplitude, and additional energy must be supplied to keep the circuit oscillating. For instance, in a vacuum tube oscillator some energy is fed back from the output to the input and returned to the output, amplified. This compensates for the positive-resistance losses. Also, a similar feedback system is used in a transistor oscillator. In both cases, a third element was used—the control grid of the vacuum tube or the base of the transistor.

The tunnel diode has no third electrode as such, but a "quantum mechanical" tunneling of energy through the diode junction. For example, while the tuned circuit (tank) is oscillating, its alternating voltage aids and opposes the DC bias across the diode, hence the diode bias increases and decreases in unison with the frequency of the tuned circuit. When the diode bias decreases, it supplies additional current in accordance with its negative-resistance characteristic; i.e., lower bias voltage, higher current.

BIASING

When a bias of .125 volt is applied to the diode, a tunnel

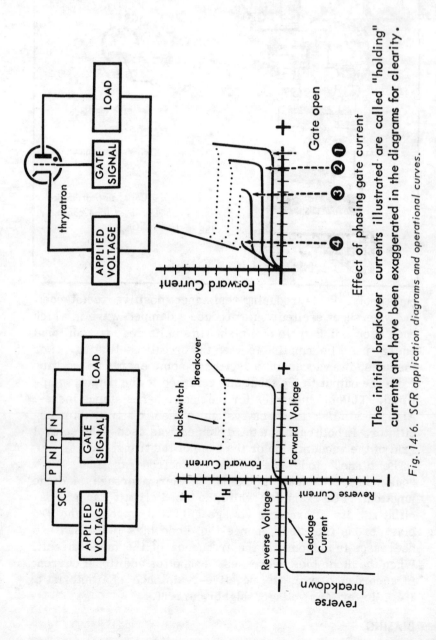

The initial breakover currents illustrated are called "holding" currents and have been exaggerated in the diagrams for clearity.

Effect of phasing gate current

Fig. 14-6. SCR application diagrams and operational curves.

current of about .5 milliampere will flow. To apply this small voltage from a 1.5-volt battery requires a series resistance of 3,000 ohms. However, the negative resistance of the diode is -100 ohms; therefore, the total series resistance is 2,900 ohms. This would obviously cancel the negative-resistance effect of the tunnel diode. To avoid this the bias voltage should be developed across a resistance value less than the negative resistance of the diode. For example, making the bias resistance 20 ohms and connecting the diode in parallel, the combined resistance will be -80 ohms. See Fig. 14-4A and B.

HALF-WAVE SILICON CONTROLLED RECTIFIER

The SCR is basically a gated, 4-element semiconductor device and its construction (NP-NP wafers) and symbol are illustrated in Fig. 14-5. In order to understand how it operates, we must first study its reverse and forward bias characteristics. Let us assume that the gate terminal is open-circuited or connected to the cathode. Connecting a small reverse bias between anode and cathode—i.e., anode negative and cathode positive—establishes a small reverse-leakage current. See Fig. 14-6. Increasing the reverse-bias voltage, a critical value is reached that causes the rectifier to break down. At this point, both voltage and current are relatively high, which causes the structure to heat beyond its rated value and, if allowed to exist, will seriously damage the rectifier. This condition must be avoided.

If we apply a forward bias between anode and cathode (gate terminal open or connected to the cathode) and gradually increase this forward-bias voltage, a point is reached where the forward current begins to rise sharply. See Fig. 14-6. Suddenly, the forward-bias voltage across the diode drops to a very small value but maintains a relatively high current. This is called the "breakover" point. However, the product of voltage and current is much smaller; therefore, the heat produced does not exceed the rated temperature of the junction as was the case with reverse biasing. This sudden switch to a low-internal voltage drop is attributed to a corresponding drop in internal resistance from a high value to a low value, hence the heat producing factors (I^2R) are held within safe limits. The rectifier, being in series with the load, will deliver maximum voltage and current to the output circuit.

The foregoing has explained the reverse and forward bias characteristics of the SCR with its gate terminal inoperative—open-circuited or shorted to cathode. However, without an active gating element, the SCR has limited application. The function of the gate electrode brings the SCR into focus as a useful control device with unlimited application. When a forward bias is applied to the gate with respect to the cathode, it provides a trigger or breakover control that can be used to advantage.

GATE SIGNAL

If we apply a small in-phase gate signal of sufficient amplitude, the positive-going gate voltage causes a forward current to flow across the gate-cathode junction. When this current reaches a critical value, it lowers the breakover point of the rectifier; at this moment, a relatively large current flows through the rectifier to the load. Once the breakover

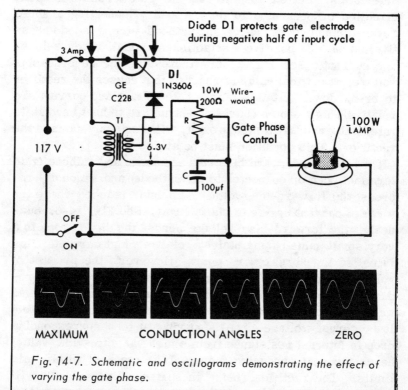

Fig. 14-7. Schematic and oscillograms demonstrating the effect of varying the gate phase.

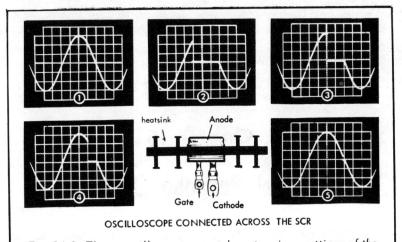

heatsink   Anode

Gate   Cathode

OSCILLOSCOPE CONNECTED ACROSS THE SCR

Fig. 14-8. These oscillograms were taken at various settings of the phase control.

has been established, the gate loses control and the rectifier continues to conduct for the remainder of the half-cycle anode voltage.

EFFECT OF GATE SIGNAL PHASE SHIFT

The circuit in the Fig. 14-7 permits shifting the phase of the gate signal with respect to the applied anode voltage. This regulates the point at which the rectifier is triggered (breakover) during each positive half-cycle. With this control, the output current can be varied to any value proportional to the conduction angle within the 180° of the applied positive half-cycle. See the oscillograms in Fig. 14-8.

Chapter 15

Special Purpose Circuits

A regulated 12-volt battery charger with a semiconductor complement is illustrated in Fig. 15-1. The system, capable of delivering 5 amperes, consists of a zener diode, a thyrector, 2 silicon rectifiers, 2 silicon controlled rectifiers, and 2 conventional diodes. Fullwave rectifiers D2 and D3 provide two positive halfwaves per cycle and supply the anode voltage to controlled rectifier SCR1, which is in series with an ammeter and battery. The positive terminal of the fullwave rectifier also supplies the forward-gate current to SCR1 via R4 and diode D5.

The battery maintains a pilot charge on capacitor C1 through potentiometer R2. When the battery is fully charged, the pilot voltage across C1 (preset by R2) is sufficient to breakdown zener diode D6. This causes C1 to discharge through D6 and R3, and the voltage developed by this resistor gates the second controlled rectifier (SCR2) into conduction. Maximum SCR2 current flows through R4 and R5, and this in turn establishes a relatively large voltage drop across R4, which opposes the gate bias of SCR1—i. e., more negative at point A. This lowers the positive SCR1 gate voltage below its cathode voltage, thereby making the gate negative with respect to its cathode. Without a forward gate bias, SCR1 switches "off" when its anode voltage falls to zero. Remember, the anode voltage consists of positive half cycles.

During the no-charge period the control circuit (which consists of R2, C1, D6, and R3) is on standby, using the pilot voltage maintained by the battery across C1. When the battery voltage drops below the D6 zener voltage, this diode is cut off and SCR2 is switched off for lack of gate bias. This,

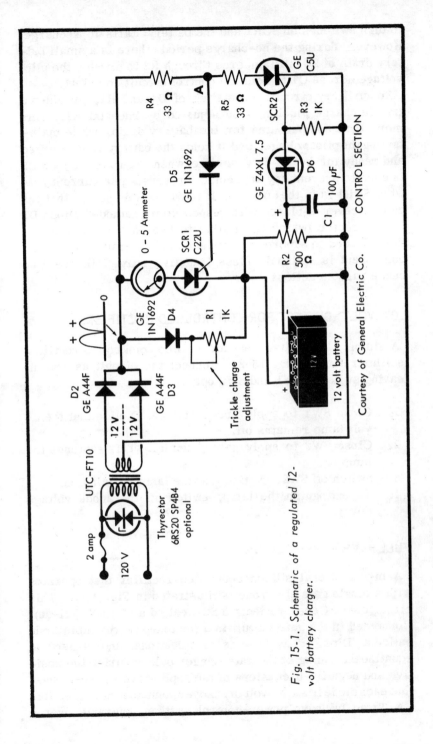

Fig. 15-1. Schematic of a regulated 12-volt battery charger.

Courtesy of General Electric Co.

in turn switches on SCR1 and the battery starts to recharge. However, during the no-charge period, there is a small battery drain of 24 milliamperes through R2 to provide the pilot voltage across C1 to hold the control circuits on alert.

An auxiliary circuit, consisting of D4 and R1, provides a trickle charge which may be adjusted by rheostat R1. This more than compensates for the battery drain, while maintaining the pilot voltage, and it holds the battery fully charged and ready for heavy-duty service when required. Diode D5 is used to prevent any appreciable reverse-gate current flow when SCR1 is in the off state. It is during this period that the gate is reverse-biased with respect to its cathode. Diode D1 is a thyrector (which is similar to two back-to-back zener diodes) used to suppress troublesome line transients. This component is optional unless inductive surges in the line cause random actions.

LOW-VOLTAGE TEST FOR CONTROLLED RECTIFIERS

A simple, low-voltage test for a silicon controlled rectifier is illustrated in Fig. 15-2. Connect the circuit as shown, leaving switches SW1 and SW2 open.

1. Close SW1 to apply anode voltage. Notice that the 3-volt lamp remains off.
2. Close SW2 to apply gate potential. This switches the lamp on.
3. Switch off SW2. Notice that the lamp remains on.
4. To extinguish the lamp, switch off the anode voltage (SW1).

FULLWAVE PHASE CONTROL

A bi-directional silicon controlled rectifier that operates with a single gate electrode is illustrated in Fig. 15-3. This circuit uses a General Electric SCR called a "Triac" (tri-ac). Connected in the gate circuit is a (matched pair) double diode called a "Diac." The Diac is bi-directional and is used to equalize the timing of the gate trigger pulses during the positive and negative excursions of the applied cycle. Tests show that each diode has a 40-volt breakover point in either direction.

A Triac fullwave power-control system, operable over a

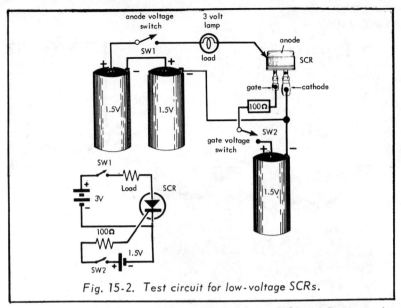

Fig. 15-2. Test circuit for low-voltage SCRs.

wide range, is illustrated in Fig. 15-4. When AC power is applied capacitor C1 in the RC circuit starts to charge, and when its voltage (VC1) reaches the Diac breakover point, it conducts and triggers the Triac. At this moment, C1 partially discharges through the Diac and the gate circuit of the Triac. After the trigger pulse the Triac conducts for the remainder of the halfwave. This occurs in both the positive and negative excursions of the applied cycle. Varying control R1 shifts the phase of the sine-wave voltage across the capacitor, which in turn shifts the phase of the Triac trigger current. Notice that the oscilloscope is connected across the load, an alternate method for measuring conduction angle. When the Diac triggers, it remains conducting for only a few microseconds.

SENSITIVE TRANSISTORIZED COUNTER

A photoelectric counter circuit in which the photocell and transistor are biased from a single battery is shown in Fig. 15-5. A Clairex photocell (CL-3) is used to trigger a 2N265 (Q1) transistor, which in turn operates a 5-milliwatt relay. A 22.5-volt battery (Burgess U15 or equivalent) supplies power to both transistor and photocell. The base bias current of Q1 is supplied from the negative side of the battery through a portion of sensitivity control R3 and fixed resistor

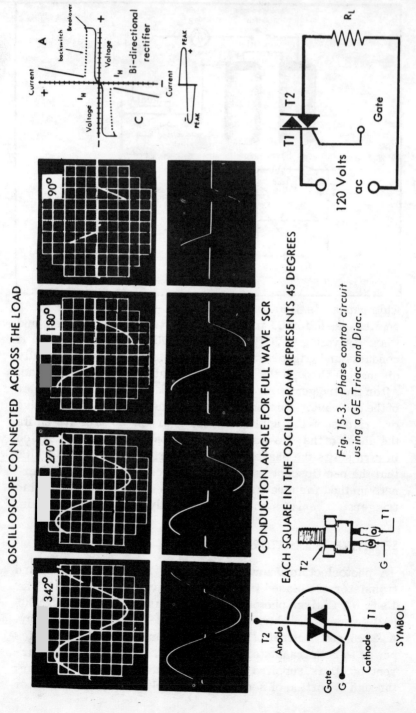

OSCILLOSCOPE CONNECTED ACROSS THE LOAD

342° 270° 180° 90°

CONDUCTION ANGLE FOR FULL WAVE SCR

EACH SQUARE IN THE OSCILLOGRAM REPRESENTS 45 DEGREES

Fig. 15-3. Phase control circuit using a GE Triac and Diac.

SYMBOL

T2 Anode
Gate G
Cathode T1

120 Volts ac

Gate

R_L

T1 T2

A
Current
backswitch
Breakover
I_H
Voltage
I_H
Voltage

C
Bi-directional rectifier
Current
+PEAK
-PEAK

208

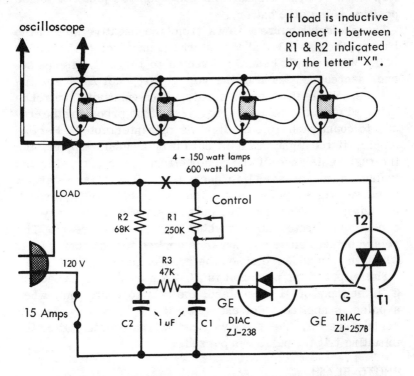

oscilloscope

If load is inductive connect it between R1 & R2 indicated by the letter "X".

4 - 150 watt lamps
600 watt load

LOAD

X

Control

R2 68K

R1 250K

R3 47K

120 V

15 Amps

C2

1 uF

C1

GE DIAC ZJ-238

GE TRIAC ZJ-257B

T2

G

T1

Fig. 15-4. Power control circuit using the Triac and Diac.

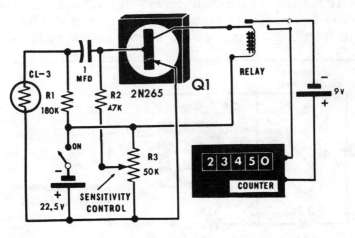

CL-3

R1 180K

1 MFD

R2 47K

2N265

Q1

RELAY

9V

ON

R3 50K

22.5V

SENSITIVITY CONTROL

23450
COUNTER

Fig. 15-5. Photoelectric counter circuit.

R2, then returns through the base-emitter junction to the positive side of the battery.

The photocell current flows from the negative side of the battery through R1 and the internal resistance of the photocell. When a light beam is directed to the cell, its resistance decreases, causing the voltage drop across R1 to increase. While this condition exists the sensitivity control is adjusted until the relay de-energizes, thus preparing the circuit to count each time the light beam is interrupted. For example, if the light beam is cut off by a passing object, the internal resistance of the photocell increases, causing the voltage drop across R1 to decrease. When this occurs a negative trigger pulse is applied to the base of Q1 through capacitor C. The negative pulse causes a momentary increase in collector current which energizes the relay. The relay contacts close and power is applied to the counter unit, which in turn adds a digit on the read-out dial.

The output of the photocell is AC-coupled through capacitor C to the input of Q1; therefore, the relay operates only when a pulse is applied. For example, if the light beam should remain cut off, the relay will cut out immediately after the initiating trigger pulse has passed.

PHOTO-FLASH

The switching device illustrated in Fig. 15-6 is shown connected for use in flash photography. The purpose of this circuit is to prevent the relatively heavy current that fires the bulb from passing through the shutter switch which would

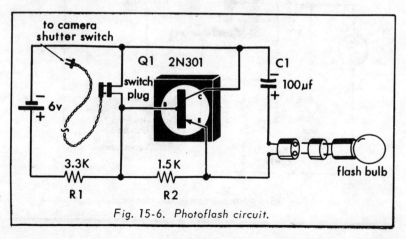

Fig. 15-6. Photoflash circuit.

eventually damage the contacts. With the shutter switch normally open, the base of Q1 is biased positively and holds the transistor at cut-off. At the same time, with the flash bulb connected, C1 (100 mfd) charges through R1, R2, and the flash bulb. The RC time constant of the charge circuit is such that the charge current is not sufficiently strong enough to fire the bulb prematurely; however, when the shutter is released the shutter switch closes and biases Q1 negatively. This causes the transistor to conduct heavily, permitting the capacitor to discharge and fire the bulb.

Notice when the shutter switch is closed that the potential across C1 (approximately 6v) drives the transistor into conduction, causing C1 to discharge rapidly through the conducting transistor and bulb. The rapid discharge is due to the very low RC time constant which permits a heavy current surge through the bulb. The transistor action permits C1 to charge through a relatively long RC time constant and discharge through a very short RC time constant. When the shutter switch opens, the capacitor will recharge when a new flash bulb is plugged in; then the circuit is ready for a repeat performance.

VOLTAGE SENSOR

As a nickel-cadmium battery discharges, a gaseous condition develops, and since the gas builds up pressure it is likely to burst the cell container if not prevented from rising beyond the critical point. Very often these batteries are used for long periods and are usually forgotten. Here, a semiconductor device is used as an electronic switch to automatically turn off the battery when its voltage drops to a prescribed level. See Fig. 15-7. When power from the battery is turned on, capacitor C1 charges through the lower portion of R3. The momentary drop across the lower portion of R3 triggers Q2 into conduction. In turn, the Q2 collector current flowing through R1 drives Q1 into saturation.

A condition now exists where the battery is discharging normally through the low resistance path of Q1 to the load. During the discharge period the battery voltage is impressed across D1 and R1. Since this voltage is higher than the zener diode level it causes a small current to flow through R3, keeping Q2 in conduction. As long as Q2 is operating, Q1 maintains conduction and the battery load circuit. However, when the battery voltage drops below the D1 zener level the diode circuit ceases to conduct and this in turn automatically switches off Q2 and Q1 in sequence, thus interrupting the battery circuit.

IMPROVED REFLEX CIRCUIT

When a vacuum tube or transistor stage functions simultaneously as a radio frequency amplifier and an audio frequency amplifier it is called a "reflex circuit." In superheterodyne receivers the last IF stage usually serves such a double-purpose. Notice in Fig. 15-8 that a shunt loop (dashed line) from the demodulator (second detector) carries back the audio component to the input of the IF stage where both the modulated IF signal and the audio frequency are combined, then amplified and separated at the output by appropriate filter circuits.

To accomplish this dual amplification function, the base input circuit of the IF stage consists of two inputs:

1. The input IF transformer (IF signal)
2. Resistance R1 (audio signal)

At the output of this stage the collector circuit contains two transformers:

1. The output IF transformer that drives the demodulator (second detector)

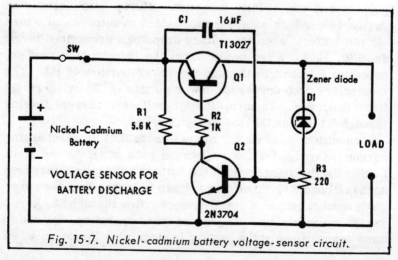

Fig. 15-7. Nickel-cadmium battery voltage-sensor circuit.

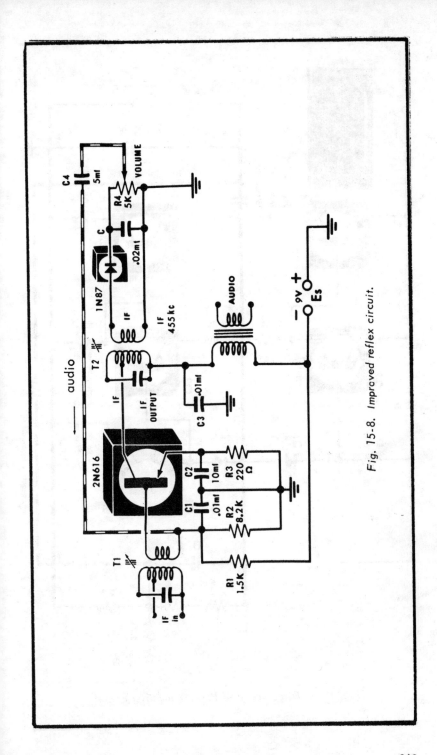

Fig. 15-8. Improved reflex circuit.

213

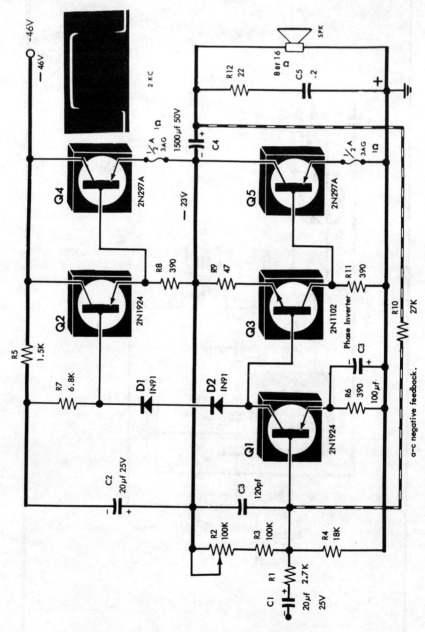

Fig. 15-9. Direct-coupled 10-watt amplifier circuit.

2. An audio transformer that drives the audio amplifier.

Since the impedance of the IF transformer is negligible at audio frequencies, the total audio output appears across audio transformer T2 while capacitor C3 bypasses the IF signal frequency to ground. This is necessary since the relatively high inductance of the audio primary would reduce the effectiveness of the IF signal. A similar bypass capacitor (C1) at the input offers a low-impedance path to ground for the IF signal.

DIRECT-COUPLED 10-WATT AMPLIFIER

High fidelity vacuum tube push-pull reproduction is somewhat of a problem using input and output transformers. Despite several methods of frequency compensation and RC phase inverter networks to eliminate the input transformer, a high quality output transformer is required for speaker coupling. Such a transformer is bulky and costly but does improve low-frequency response. An attempted alternative was to eliminate the output transformer by using a bank of vacuum tubes in parallel capable of delivering a high current direct to the speaker voice coil.

When power transistors became available, circuit design engineers turned their attention to these low-voltage, high-current devices and developed a transformerless hi-fi power amplifier. Such a circuit using five transistors is illustrated in Fig. 15-9. Input stage Q1 is a Class A driver with an emitter current of about 3 milliamperes. The forward base bias is obtained from voltage divider network R2, R3, and R4. Variable resistance R2 is used to adjust the bias so that 23 volts (half the supply) appears across Q5.

The RC circuit in the Q1 emitter lead provides DC stabilization. R2 and R3 connected to the base of Q1 provide negative feedback, and shunting these two resistors is capacitor C3 whose value is selected to provide a good 2-kHz square-wave response at the speaker terminals. An additional negative feedback loop is connected from the speaker output to the base of Q1 to further improve amplifier performance. See dashed line in Fig. 15-9.

The collector of Q1 drives (simultaneously) Q2, an emitter-follower, and Q3, an NPN phase inverter. Both Q2 and Q3 are designed for Class B push-pull operation. The forward bias current in these two transistors is approximately 1 milli-ampere to reduce crossover distortion. The bias is supplied by the voltage drop across two temperature-sensitive germanium diodes, D1 and D2. Each diode has a negative-resistance temperature coefficient similar to the emitter-base junction of the two germanium transistors, Q2 and Q3. For example, an increase in temperature decreases the resistance of the base-emitter junction, resulting in an increase of base current. However, the resistance of each diode also decreases and lowers the base bias voltage. The decrease of forward bias is about 2 millivolts per degree Centigrade.

Transistors Q4 and Q5 operate with a forward bias of approximately 15 milliamperes supplied by the voltage drop across R8 and R11, respectively. Notice that R8 shunts the input of Q4. Resistor R9 (47 ohms), connected in the Q3 emitter lead, is used as a stabilizer and local feedback. Transistor pairs, Q2-Q4 and Q3-Q5, are operated in Darlington connection to increase current gain. The .5-ampere fuse in each emitter lead is used for protection and, since each fuse is about 1-ohm resistance, they serve a dual purpose of local negative feedback (AC) for frequency response and DC negative feedback for stability.

The speaker (voice coil) coupling capacitor C4 is a 1500-microfarad electrolytic that charges and discharges via the voice coil. For example, when Q4 is driven, C4 charges and the charge path is from the negative supply terminal through Q4 (collector-emitter), C4, and the voice coil. On the alternate swing of the signal, Q5 is driven and C4 discharges through Q5. This makes the amplifier a single-ended operation with all the advantages of Class B push-pull, and since the voice coil is driven by the Q4 emitter current, it qualifies as an emitter-follower with low-inherent distortion and low-output impedance. Therefore, from the output of Q1 to the output of Q4 (voice coil), it is push-pull emitter-follower operation. See Fig. 15-10 for the basic circuit. The output impedance to voice coil is approximately 1 ohm which provides good speaker damping.

A positive feedback voltage appears via C2 across R5, providing bootstrap action to Q2. This boosts the positive swing of the output signal to equal the negative swing and corrects an unsymmetrical condition in the output circuit. This positive feedback is balanced out by the negative feedback of R2

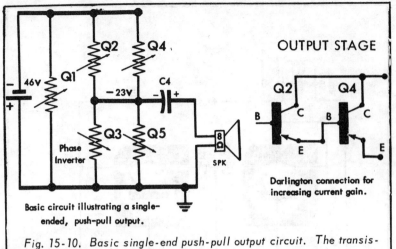

Basic circuit illustrating a single-ended, push-pull output.

OUTPUT STAGE

Darlington connection for increasing current gain.

Fig. 15-10. *Basic single-end push-pull output circuit. The transistor symbols illustrate the Darlington connection.*

and R3 at the input of Q1. However, due to the feedback from output to input through R10, the overall negative feedback on the system predominates. The series RC circuit shunting the voice coil offsets the continuous rise of load impedance and phase shift at high frequencies. The amplifier will deliver 8 watts of continuous output power with a 1-volt RMS input signal or 10 watts of program material. Power transistors Q4 and Q5 are mounted on a heat sink.

TRANSFORMERLESS PUSH-PULL AUDIO SYSTEM

A typical complementary symmetry audio system appearing in some transistor radio receivers is illustrated in Fig. 15-11. The circuit consists of a driver Q1 and a Class B push-pull stage Q2 and Q3. When a negative-going signal is applied to the base of the input stage (PNP) its collector current (flowing through R3) increases and develops a positive voltage at the base of transistors Q2 and Q3. Since Q2 is a PNP unit, the positive base bias causes this stage to approach cutoff and Q3 (PNP) to conduct.

While this bias condition exists, an electron current flows from the negative terminal of the battery (Es) through the speaker voice coil, and returns to the positive terminal of Es via R3, D, and Q1, which is conducting. At the same time, a second current flows from the negative terminal of Es and charges C1, then returns to the positive terminal of

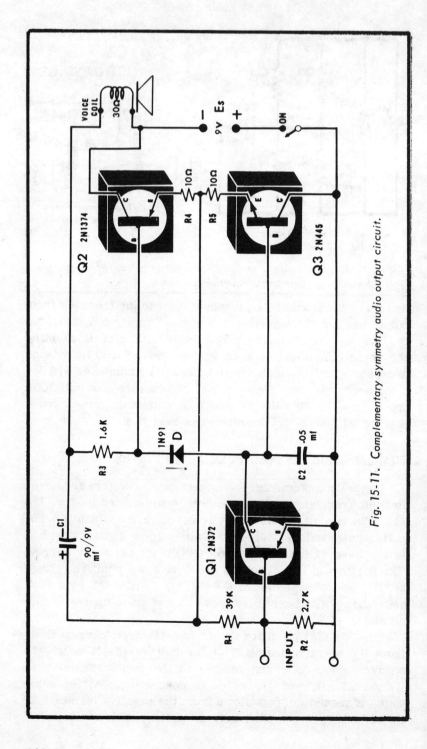

Fig. 15-11. Complementary symmetry audio output circuit.

Es via R5 and conducting transistor Q3. Notice the C1 charge current passes through the voice coil, which reproduces the negative half of the input signal. When a positive-going signal is applied to the base of input stage Q1, its collector current (through R3) decreases and develops a negative-going voltage at the base of transistors Q2 and Q3. Since Q3 is an PNP unit the negative base bias causes this stage to approach cut-off and Q2 (PNP) to conduct.

During this condition, C1 discharges through the voice coil, conducting transistor Q2, and returns via R4 to the positive side of C1. This time, however, due to the discharge of C1 the electron current flows in the opposite direction to that of the charge current and reproduces the positive half of the input signal. This completes one cycle of the applied audio signal. The diode, connected between the base of transistors Q2 and Q3, functions as a base bias regulator and minimizes crossover distortion by maintaining the correct base bias threshold for each transistor.

PHONO PREAMPLIFIER

An audio preamplifier using a balanced feedback network from collector to base appears in Fig. 15-12. It was designed for use with a ceramic cartridge and two G.E. 2N2926 silicon transistors. Capacitor values should be within plus or minus 10% of the suggested values. The total current drain is 2 milliamperes at 22 volts. The output is taken from an emitter-follower stage which functions as a buffer between the preamplifier and subsequent equipment.

LOW-FREQUENCY COMPENSATION NETWORKS

The frequency-compensation circuits in Fig. 15-13 consist of two RC networks—one; the coupling between the output of Q1 and the input of Q2, and second; the coupling between the output of Q2 and the input of Q3. Capacitor C1 (10 mfd) functions as a coupling capacitor while the smaller unit C2 (1 mfd) functions as an input drive capacitor to Q2. Since the impedance of C2 is relatively high at low frequencies and low at high frequencies, the drive voltage to the input of Q2 follows the same order; however, to prevent serious loss in drive at the high-frequency end of the audio range, a resistance (R5) is connected in series with C2. This, of course, increases the total drive impedance at all frequencies; however, the low-frequency drive still predominates.

The second low-frequency compensating circuit between Q2 and Q3 performs the same function. The value of the series resistance is important in each case and should be a compromise between high-frequency audio and noise. This type of low-frequency compensation is similar in principle to basic

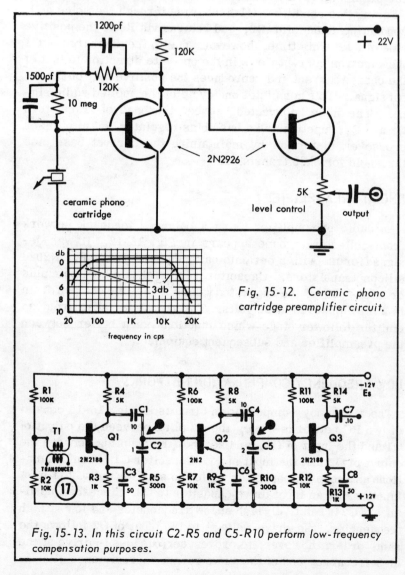

Fig. 15-12. Ceramic phono cartridge preamplifier circuit.

Fig. 15-13. In this circuit C2-R5 and C5-R10 perform low-frequency compensation purposes.

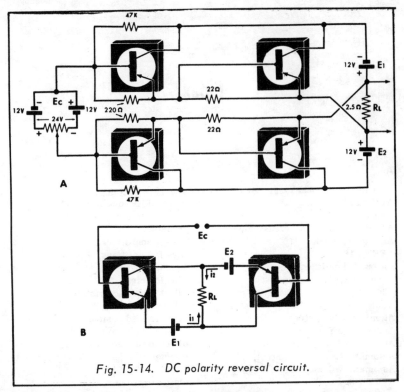

Fig. 15-14. DC polarity reversal circuit.

tone control circuits used in radio. However, in tone control the series resistance is made variable to function as a control.

POLARITY REVERSAL CIRCUIT

The basic principle of this circuit is illustrated in Fig. 15-14. Notice that the emitter-collector of each transistor is common to R_L and connected so that their currents oppose one another. The input or control voltage Ec is DC and is applied to the base of each transistor, causing one transistor to cut off and the other to conduct. When the control voltage is reversed, the conduction and cut-off condition switches from one transistor to the other causing the current through R_L to reverse. The voltage gain is almost unity; current gain is approximately Beta x R_{L_1}.

Index